Advances in Soil Science

Advances in Soil Science

B.A. Stewart, Editor

Advances in Soil Science

Volume 19
Limitations to Plant Root Growth

Edited by J.L. Hatfield and B.A. Stewart

With Contributions by
J.M. Baker, F.N. Dalton, C.D. Foy, B. Klepper,
J.J. Meisinger, J.F. Power, A.N. Sharpley,
D.L. Suarez, W.B. Voorhees, J.M. Wraith, R.W. Zobel

With 41 Illustrations

Springer-Verlag
New York Berlin Heidelberg London Paris
Tokyo Hong Kong Barcelona Budapest

J.L. Hatfield
National Soil Tilth Laboratory
USDA Agricultural Research Service
Ames, IA 50011 USA

B.A. Stewart
USDA Conservation and Production Research
Laboratory
Bushland, TX 79012 USA

ISSN: 0176-9340

Printed on acid-free paper.

Production managed by Ellen Seham; manufacturing supervised by Genieve Shaw.
Typeset by Best-set Typesetter Ltd., Chaiwan, Hong Kong.
Printed and bound by Edwards Brothers, Inc., Ann Arbor, MI.
Printed in the United States of America.

9 8 7 6 5 4 3 2 1

ISBN 0-387-97767-8 Springer-Verlag New York Berlin Heidelberg
ISBN 3-540-97767-8 Springer-Verlag Berlin Heidelberg New York

Preface

In the past 10 years, there has been a resurgence in interest in soil management and conservation of the soil resource. With the knowledge we have accumulated in the past 100 years, there is the possibility of developing new and innovative ways of effectively managing the soil. The emphasis on sustainable agriculture requires that we understand how to utilize the soil as a viable living resource. To meet the world demand for food within the next 50 years requires a healthy and strong soil resource which can sustain production.

With the dedication of the National Soil Tilth Laboratory in 1989, it was decided that one contribution the laboratory could make to agricultural science would be to foster an exchange of information on soil management. The focus of that interchange centers on long-term soil management. If we are to fulfill the goals of sustainable agriculture, environmental quality, and feeding the world, there will have to be an increased understanding of how to effectively manage the soil. Long-term soil management requires integrated and interdisciplinary research to bring all of the information together in terms which would be applicable to all soils. To accomplish this goal a workshop is held each year, with the exchange of information focusing on a single topic within the framework of developing effective strategies for long-term soil management.

With the forum to focus on an individual theme each year, the theme for the initial workshop was "Limitations to Plant Root Growth." This topic was chosen because it was felt that, before we could understand how to manage the soil, limitations to plant root growth should be overcome. The speakers for this workshop were invited to present their ideas about their perspectives on the individual topics which are represented within this volume. We included soil compaction as an additional topic, since there are large amounts of information in the literature that have not been included in this volume. There are many individuals who were not able attend this workshop and they will benefit from having this information available. The chapters contain a wealth of information on different aspects of root growth. There are many topics which could have been included; however, the authors have presented an excellent set of papers and a thought-provoking set of needed research. We hope that all researchers will consider these topics worthy of research and will assist in finding the answers to these questions.

The chapters show that there is much we understand about the plant root system; however, there are many areas in which we need to gather more information. To fill these knowledge gaps, there needs to be a concerted effort to create teams which would combine several disciplines together. The goal of these workshops is to promote these interdisciplinary teams with the long-term goal of developing an understanding of how best to manage and improve the soil.

J.L. Hatfield
B.A. Stewart

Contents

Preface v
Contributors ix

Development and Growth of Crop Root System 1
Betty Klepper
I. Introduction 1
II. Monocotyledonous Root System 2
III. Dicotyledonous Root System 9
IV. Needed Research 14
References 18

Soil Environment Constraints to Root Growth 27
Richard W. Zobel
I. Introduction 27
II. The Rhizosphere Environment 29
III. Root System Morphology 36
IV. Root-Soil Interactions 37
V. Conclusions 46
References 48

Root Function in Water Transport 53
J.M. Baker, J.M. Wraith, and F.N. Dalton
I. Introduction 53
II. Water Entry into Roots 54
III. Root Function During Drought 63
IV. Concluding Remarks 66
References 67

Wheel-Induced Soil Physical Limitations to Root Growth 73
Ward B. Voorhees
I. Introduction 73
II. Soil Physical Limitations to Root Growth 74
III. Extent and Influence of Wheel Traffic 75
IV. Root Growth 80
V. Root Growth and Crop Yield 85
VI. Subsoil Compaction 88
VII. Conclusions 91
References 92

Soil Chemical Factors Limiting Plant Root Growth 97
Charles D. Foy
I. Introduction 97
II. Soil Chemical Factors 98
III. Solving the Shallow Rooting Problem 116
IV. Challenges for Future Research 129
V. Summary 130
VI. Acknowledgments 131
References 131

Root Extraction of Nutrients Associated with Long-Term Soil Management 151
A.N. Sharpley, J.J. Meisinger, J.F. Power, and D.L. Suarez
I. Introduction 151
II. Effects of Long-Term Soil Management on Root Extraction of Nutrients 152
III. Interaction Among Management Factors 189
IV. Modeling Long-Term Management Effects on Root Extraction of Nutrients 192
V. Conclusions 199
References 200

Index 219

Contributors

J.M. Baker, U.S. Department of Agriculture, Agricultural Research Service, Soil and Water Management, St. Paul, MN 55108, USA

F.N. Dalton, U.S. Department of Agriculture, Agricultural Research Service, U.S. Salinity Laboratory, Riverside, CA 92501, USA

Charles D. Foy, U.S. Department of Agriculture, Agricultural Research Service, Climate Stress Laboratory, National Resources Institute, Beltsville, MD 20705, USA

Betty Klepper, U.S. Department of Agriculture, Agricultural Research Service, Columbia Plateau Conservation Research Center, P.O. Box 370, Pendleton, OR 97801, USA

J.J. Meisinger, U.S. Department of Agriculture, Agricultural Research Service, Beltsville, MD 20705, USA

J.F. Power, U.S. Department of Agriculture, Agricultural Research Service, Department of Agronomy, University of Nebraska, Lincoln, NB 68583, USA

Andrew N. Sharpley, U.S. Department of Agriculture, Agricultural Research Service, National Agricultural Water Quality Laboratory, Durant, OK 74702, USA

D.L. Suarez, U.S. Department of Agriculture, Agricultural Research Service, U.S. Salinity Laboratory, Riverside, CA 92521, USA

Ward B. Voorhees, U.S. Department of Agriculture, Agricultural Research Service, North Central Soil Conservation Research Laboratory, Morris, MN 56267, USA

J.M. Wraith, U.S. Department of Agriculture, Agricultural Research Service, Soil and Water Management, St. Paul, MN 55108, USA

Richard W. Zobel, U.S. Department of Agriculture, Agricultural Research Service, 1017 Bradfield Hall, Cornell University, Ithaca, NY 14853-1901, USA

Development and Growth of Crop Root Systems

Betty Klepper

I. Introduction 1
II. Monocotyledonous Root System 3
III. Dicotyledonous Root System 9
IV. Needed Research 14
References 18

I. Introduction

All crops grown from seed initiate root establishment with the emergence of the radicle from each germinating seed. The radicle anchors the seedling for the upward thrust needed in the emergence process and also collects the water required to generate turgor pressure needed for elongation of the emerging shoot. This first root on the plant may become a permanent and important component of the root system in taprooted plants such as cotton (*Gossypium hirsutum* L.) or it may atrophy or comprise a relatively insignificant fraction of the root system, as it does in cereals like corn (*Zea mays* L.). In either case, radicle growth, the first step in plant establishment, is especially important because of its role in water absorption during emergence and early stand establishment.

The events which follow radicle elongation differ between monocotyledonous and dicotyledonous species. These events are also modified by environmental factors such as soil temperature, fertility, aeration, and strength and by plant factors such as seedling vigor, especially the ability of the shoot to supply carbohydrates and other nutrients to developing root parts.

Advances in Soil Science, Volume 19

This chapter describes the developmental history of monocot and dicot root systems with special emphasis on the timing and sequencing of the appearance of morphological features in the root system, including reference to the effects of environmental and endogenous factors on this development. Root growth (elongation or dry weight increase) will also be discussed, primarily in relation to dry matter distribution in crops. Finally, areas in need of further research will be addressed, especially with respect to the interactions between developing root systems and soil properties.

II. Monocotyledonous Root Systems

A. Axis Production, Elongation, and Direction of Growth

Monocot root systems are comprised of many root axes and their branches. These root systems are generally called "fibrous" roots because of the massive body of small wiry roots in the top 10 to 20 centimeters of soil. These root axes originate from nodes either in the seed (seminal roots) or in the crown (adventitious or coronal roots). Each crown node passes through a series of developmental events including leaf elongation, tiller production, and root axis production. Control of the time of appearance of individual axes from each node is related to shoot development (Klepper et al., 1984). This relationship has been worked out for sorghum (*Sorghum bicolor*) (Mirhadi and Kobayashi, 1980), rice (*Oryza sativa*) (Yoshida et al., 1982), maize (*Zea mays*) (Picard et al., 1985), and wheat (*Triticum aestivum*) (Rickman et al., 1985). Generally, the seminal root system services the shoot until the crown root system is produced beginning at about the three leaf stage (Klepper et al., 1984; Newman and Moser, 1988). The number of axes produced at each node varies, and more root primordia may be developed than ever elongate to produce visible roots (Girardin et al., 1987). Lower nodes generally have 2 to 4 axes; upper nodes may produce more axes, some of which may assist in propping up the crop against lodging (Gregory, 1983; Picard et al., 1985). Nodes on tillers also generate axes and earlier tillers have older, deeper root systems than younger ones (Yoshida et al., 1982; Anderson-Taylor and Marshall, 1983; Rickman et al., 1985). The timing of axis appearance can be predicted either by reference to leaf number on shoots (Jordan, 1987) or by use of cumulative thermal time (growing degree days, GDD), which is exponentially related to the number of nodal axes per plant in wheat (Vincent and Gregory, 1989a,b).

The wheat seminal root system elongates directly downward beneath the plant and axes grow more or less vertically; the crown roots grow outward at an angle for a time before turning and growing vertically downward (Klepper, 1990). Thus, axes from higher nodes in the crown explore a widening cylinder of soil around the cylinder directly beneath the plant already filled by the seminal root system. Factors controlling this behavior of root axes are not known, but Mosher and Miller (1972) found that soil temperature controlled the direction of growth of root axes of maize, indicating that temperature may be one of the factors involved.

The rate of exploration of soil depth by these elongating axes is controlled primarily by plant vigor and by soil environmental conditions, especially temperature, moisture, and strength. Warmer soil and larger, older plants produce axes which elongate rapidly under greenhouse conditions (Gregory, 1986). Under field conditions, rates of axis elongation are of the order of one to several centimeters per day (Kaigama et al., 1977). In a greenhouse experiment where soil temperature was controlled at 31°C, Gregory (1986) found that the rates of axis elongation for pearl millet (*Pennisetum typhoides* S. & H. cv BK560) ranged from $1.4\,\mathrm{cm\,d^{-1}}$ to $7.9\,\mathrm{cm\,d^{-1}}$. A similar range was found in studies at the Auburn rhizotron for a variety of crops (Taylor et al., 1970). Field rates may be severely restricted by compaction, acid subsoils, or other inhospitable conditions which may also cause vertically-oriented axes to be reoriented horizontally or to be stopped altogether.

B. Lateral Branching

Each axis produces lateral root primordia at a set point in its developmental history. Generally, a single axis can be viewed as being made up of a series of age zones with the youngest tissue occurring immediately behind the tip in the elongation zone. Slightly older is the zone of maturation where root hairs are visible on the root exterior. At the upper (older) end of the root hair zone, one can find, in the pericycle of the stele, primordia for potential lateral roots which grow through the endodermis, cortex, and epidermis to break out of the parent root, usually several centimeters behind the tip (Klepper, 1990). While axes can be considered the primary agents for extending root depth, the laterals are the principal agents for increasing root length density at a given depth. The first-order laterals produced from vertically-growing axes grow at right angles to the parent axis and, in like manner, the second-order laterals grow at right angles to their parent roots. The end result of these growth habits is an orthogonally-shaped root system.

The idealized root system is exponentially distributed throughout its depth with many laterals and their branches in the upper part of the profile and a high proportion of axes with fewer branches in the lower part of the profile. This type of distribution is common in moist rain-fed and irrigated conditions (Gerwitz and Page, 1974; Barraclough and Leigh, 1984), but fails to describe root distribution in soils with dry surface layers or other inhospitable horizons (Klepper, 1991).

Approximately 250 GDD (base temperature 0°C) are required for newly expanded wheat roots to produce visible lateral roots (Klepper et al., 1984). Assuming 100 GDD per leaf, this represents 2.5 phyllochrons. In sorghum, Mirhadi and Kobayashi (1980) found about two phyllochrons elapsed between the first visible appearance of primordia and the elongation of lateral roots.

C. Anatomical Features

Although monocot roots do not increase in diameter from secondary growth, they do undergo development to form features such as endodermal thickening, cortical aerenchyma, cortical senescence, and lateral branching. These features are important in root function and have been studied closely for some time.

The endodermis is a cylinder of tissue comprising the outermost cell layer of the stele. At the least, it is surrounded by several layers of cortical cells and an epidermal layer. All materials which traverse the root must pass through this cell layer, which apparently has a high resistance to movement of water and ions (Robards and Robb, 1972; Clarkson and Sanderson, 1974). In a detailed study of barley (*Hordeum vulgare*), Robards et al. (1973) described three phases in endodermal cell development, including an early phase with only Casparian bands in anticlinal walls, a second phase in which the suberin lamella surrounds the inner face of the wall, and a third phase with cellulose deposits thickening the wall. Other monocots have similar features (Fritsch, 1977; Clark and Harris, 1981). The endodermis affords a major resistance to transport from the root surface to the root xylem because suberization of the walls forces all materials passed across the endodermis to go through the symplasm rather than through cell walls. Resistances of the order of $2 \times 10^{-7}\,\mathrm{m\,s^{-1}\,MPa^{-1}}$ have been measured for water transport across root tissues (Klepper, 1990).

In the absence of significant secondary growth, lateral branches of monocot roots must be joined onto the already mature xylem in the stele. In rice, protoxylem elements in the daughter root are connected with protoxylem or early metaxylem elements of the parent root, and the late metaxylem elements of the daughter are connected with late metaxylem elements of the parent root. This connection is made by differentiation of parenchyma cells into tracheary elements (Kawata et al., 1977). Water and minerals collected by lateral roots must pass through these bridging cells into the xylem of the parent root and these "joins" may be sites of resistance to flow (Klepper, 1983).

The location of the lateral roots on the parent root is not random. In the onion (*Allium cepa*) lateral roots occur in vertical files opposite the protoxylem poles with one or two of the poles predominating (Pulgarin et al., 1988).

D. Numbers of Roots

The timing of the appearance of morphological units making up the root system is a feature of plant development; the numbers of roots, their length, and their dry weights are all features of plant growth. In general, the better the growing conditions, the higher the fraction of potential roots that will appear. For example, fewer root axes appear at low irradiances compared to high irradiances in winter wheat (Vincent and Gregory, 1989a). In rice, the total number of crown root primordia associated with the upper nodes is decreased by low nitrogen supply and low-light intensity (Kawata et al., 1978). Plant vigor is an important factor in determining the number of axes produced in maize (Jordan, 1987) and, consequently, year-to-year variation in the numbers of axes at any one node (Picard et al., 1985) is not unexpected. Nevertheless, in a field experiment where plants were not stressed, the numbers of nodal axes of wheat (y) were exponentially related to thermal time (x) by the equation, $\ln y = 0.0056x - 2.324$, $r^2 = 0.96$ (Vincent and Gregory, 1989a,b). One would expect the number observed under stress conditions to fall beneath this line.

The number of lateral roots elongated on parent roots (branching frequency, number mm^{-1}) varies with local soil fertility under both laboratory (Drew and Saker, 1975) and field (Belford et al., 1987) conditions where nitrogen fertilizers are applied. In a greenhouse experiment on barley, May et al. (1965) found that plants supplied with a full nutrient solution produced the same numbers of root axes as those provided with 10% nutrient, but the root systems in full nutrient had fewer first- and second-order laterals than those in the diluted nutrient. On the other hand, Feil and Geisler (1988), working in greenhouse trials with winter wheat, found that increasing the nitrogen supply in the nutrient solution to 900 $mg\,L^{-1}$ N from 11 $mg\,L^{-1}$ N increased the numbers of laterals.

When sorghum plants were hydroponically grown under adequate nutrient supplies but with the root axis number restricted to three, branching of second- and third-order laterals was promoted compared to that of control plants with 6 or 10 axes, indicating that supplies of photosynthate or other shoot-provided materials may limit branching frequency in intact plants (Jordan et al., 1979).

Soil water can also impact the appearance of lateral roots. For example, Jupp and Newman (1987) found that the branching frequency on axes of *Lolium perenne* L. (perennial ryegrass) grown in soil increased during a drying cycle until soil matric potential fell below about -10 MPa, when new laterals ceased appearing and existing root tips began to die off. As the soil was drying down during the cycle, rates of lateral root producting were increased to around 17 roots $m^{-1}\,d^{-1}$ when compared to values for well-watered controls of about 3 roots $m^{-1}\,d^{-1}$. With rewetting, previously existing lateral roots elongated and production of new laterals was suppressed.

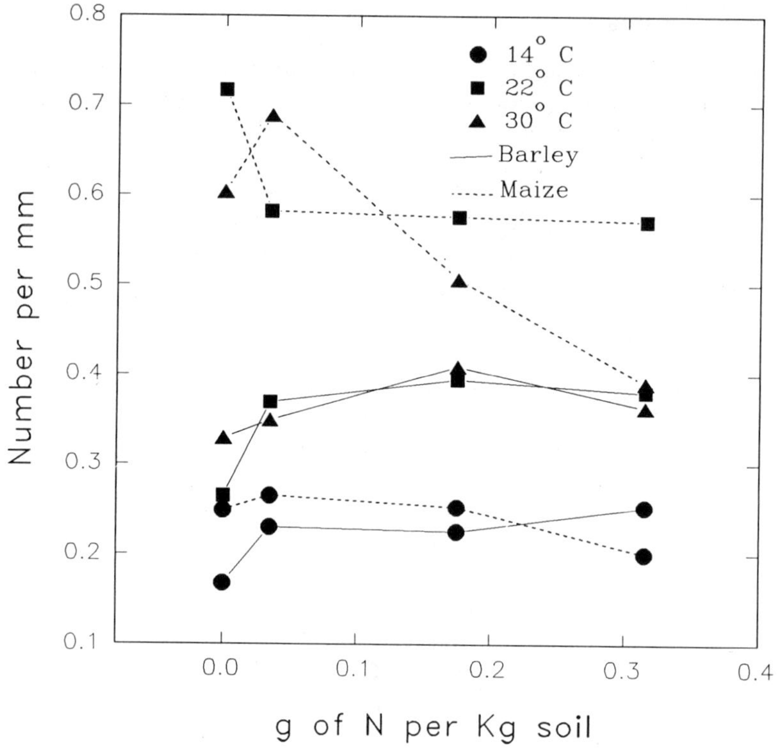

Figure 1. Plotted with data from Geisler and Krützfeldt, 1983.

Geisler and Krützfeldt (1983) did a detailed greenhouse and growth chamber study on the effects of temperature, nitrogen fertilization, cultivar, and plant age on root morphology of spring barley, maize, and field bean (*Vicia faba* L.) grown in a sand-soil mixture. In general, branching frequency in both barley and maize was greater for 15-day-old plants than for 5-day-old plants. Cultivar differences were found but were not especially notable. Figure 1 illustrates the effects of temperature and nitrogen on the branching frequency for these two species. This figure illustrates how different investigators could find quite different trends in studies done under different temperature or fertility conditions. If 14°C had been the test temperature, for example, the investigators would have found very little impact of fertility on branching frequency and very little difference between the two species. With a test at 22°C, conclusions would have been that fertility increased lateral rooting in barley and decreased it in maize and that maize had about twice as many laterals per millimeter of axis as barley. Taking the temperatures most likely to be encountered by each species as 14°C for spring barley and 22°C for the warm season maize, one could conclude that, under normal field conditions, elevated nitrogen fertility would probably increase the lateral root number per millimeter of axis in barley and decrease it in maize and that maize produced more than twice the number of laterals per unit length of axis as barley.

E. Lengths, Diameters, and Weights of Roots

Generally, the order of branching determines the length and diameter as well as the weight per unit length (lineal density) of monocot roots. Axes are usually the longest individual roots (up to two or more meters) and the largest in diameter and have the highest lineal density; first-order branches are intermediate in these properties; and higher-order branches are the smallest. However, contributions of these different orders of branching to the total root length occur in a different pattern. For example, after eight weeks of growth of barley in a controlled environment chamber, Walters and Ayres (1981) found nine coronal axes with an average length of 47 cm (total length, 425 cm); 715 first-order laterals appeared on the crown root system with an average length of 1.75 cm (total length, 1251 cm); and 207 second-order laterals appeared with an average length of 0.15 cm (total length, 29 cm). In this experiment, the 1705 centimeter-long nodal root system was comprised of 25% axis, 73% first-order laterals, and about 2% second-order laterals. On these same plants, the seminal root system was 2917 cm long and consisted of the same percentages of the different root types as the nodal system. There were seven seminal axes with an average length of 105 cm (total length, 732 cm), 1782 first-order laterals with an average length of 1.2 cm (total length, 2132 cm), and 489 second-order laterals with an average length of 0.1 cm (total length, 53 cm). Thus, the seminal and nodal root systems in this experiment were very similar in gross morphology with 73% of the length coming from first-order laterals. Nutrient deficiencies decreased the average length of the laterals. Deficiencies of phosphorus and potassium significantly reduced the lateral length in two cultivars of barley (Hackett and Bartlett, 1971). Nitrogen also increases the length of the laterals in wheat (Belford et al., 1987) as well as their diameter (Hackett, 1972).

The diameter and lineal density of monocot roots are interrelated. In a study of two perennial grasses, Cornish et al. (1984) found that the lineal densities of both ryegrass (*Lolium perenne* L. cv. Rvanui) and phalaris (*Phalaris aquatica* L. cv. Australian) were reduced by a factor of about three under conditions where high axial resistances induced shoot water stresses. For example, values for ryegrass ranged from $3.3\,mg\,m^{-1}$ to $0.8\,mg\,m^{-1}$ and for phalaris from $4.5\,mg\,m^{-1}$ to $1.4\,mg\,m^{-1}$. Macduff et al. (1986) found that temperature affected radius and linear density of barley roots growing for 20 days in a solution culture held at different temperatures. Under warm temperatures (25°C), the average root radius was 0.11 mm and the lineal density was $3\,mg\,m^{-1}$; at 3–7°C, the radius was 0.13 mm and the lineal density $10\,mg\,m^{-1}$.

Mineral nutrients can also affect the diameter of lateral roots. For example, Christie (1975) found that phosphorus deficiency for three grasses decreased the average root diameter to 85% of its value on normal plants. For example, mulga grass root radius was 0.20 mm at $3\,mg\,L^{-1}$ phosphorus and 0.17 mm at $0.003\,mg\,L^{-1}$ phosphorus. This probably resulted, in part, from the mixture of root classes making up the average with more nodal roots present on the plants grown at $3\,mg\,L^{-1}$ phosphorus.

F. Dry Matter Partitioning

The general pattern of dry matter distribution over the life of the crop shows a shift from a heavy investment in roots during seedling establishment and early vegetative growth in the first part of the growing season to a heavy investment in reproductive structures during the latter part of the season (Klepper, 1991). For example, Anderson-Taylor and Marshall (1983) found that the root to shoot ratio of spring barley (*Hordeum distichum* L. Lam.) decreased from 0.3 one week after planting to 0.1 twelve weeks after planting. In wheat, between the double-ridge stage and anthesis, dry matter accumulated into the shoots ten times faster than into the roots, resulting in a root to shoot ratio of about 0.1 at anthesis compared to 0.7 at early tillering (Barraclough, 1984).

In addition to the strong seasonal effect on dry matter partitioning patterns, stresses can also influence these patterns. Generally, high root to shoot ratios are positively correlated to stress. Unfavorable temperatures and water stress tend to generate high root to shoot ratios (Geisler and Krützfeldt, 1984; Klepper, 1991). On the other hand, high fertility generally has the opposite effect (Ennik and Hofman, 1983).

G. Modeling Concepts for Monocot Root Systems

Cereal root systems can be mathematically simulated using rather simple concepts (Klepper and Rickman, 1990, 1991). These concepts include the sequential production of downward-growing axes with time; the elongation of these axes through the rooting medium at a rate determined by endogenous and environmental factors; the production of branches along these axes at a time and lineal frequency controlled by plant and soil factors; and the elongation and branching of these laterals by a system of rules similar to those for axis elongation and branching. Also, root to shoot partitioning rules are needed so that carbohydrate limitations can be quantified.

Many of these developmental concepts are incorporated into a model, WHTROOT (Porter et al., 1986). In this model, growing degree days

(GDD) are calculated for each 10-cm depth increment in the profile. Root elongation is assumed to be linearly related to GDD. Axis production is controlled by the GDD of the top increment. Branching is assumed to occur in each depth increment for each segment of root 250 GDD after the root segment first appears. Axes grow vertically downward and branches grow at right angles to their parent root. The model does not include restrictions associated with adverse soil or plant conditions. Vertical axes, first-order laterals, and higher-order laterals are assumed to elongate at different rates and to have different diameters and lineal densities. This model has been applied to soil with different levels of compaction (Klepper and Rickman, 1990). Rickman et al. (1992) used a soil elastic modulus to describe restrictions on root elongation and used this same model to show that root length densities predicted were in the general patterns observed in the field.

Partitioning rules have been developed for grasses. For example, Johnson and Thornley (1983) allocated 10 percent of the available carbohydrate to roots in an established grass sward. Skiles et al. (1982) assumed that sink strength is proportional to sink size and provided that root respiration rate and mortality depend on temperature and water stress. A number of other rules were described by Klepper and Rickman (1991).

III. Dicotyledonous Root System

A. Taproot Elongation, Direction of Growth, and Secondary Thickening

Dicot root systems can take a variety of forms. The form which most people think of first is the strongly taprooted system where the entire root system originates from the primary root or radicle. This root grows vertically and has a system of subordinate branches that explore soil to the side. These taprooted plants include fleshy storage roots such as carrots (*Daucus carota* L.) and woody roots such as those found on trees and shrubs and in such annual crops as cotton (*Gossypium hirsutum* L.). Some dicots have modified taprooted systems with a weaker taproot and stronger first-order laterals so that a more diffuse or fibrous root system results. Modified taproots are common in such crops of the legume family as beans (*Phaseolus vulgaris* L.) and peas (*Pisum sativum* L.) where the radicle atrophies after seedling establishment and a flush of laterals is produced with no particular lateral assuming dominance. Soybeans (*Glycine max* L. Merr.) are diffuse-rooted, but the taproot elongation rate initially determines rooting depth and some cultivars have a more dominant taproot than others (Kaspar et al., 1984). White clover cultivars also show a range of relative dominance of taproots (Caradus, 1981) as do alfalfa cultivars (McIntosh and Miller, 1980). For all species, taproot

Table 1. Taproot elongation rates of several dicot species.

Crop species	Test situation	Medium	$cm \cdot d^{-1}$ rate	Reference
Soybean (*Glycine max* L. Merr)	Plastic tubes, greenhouse	Vermiculite	3.0–5.3	Kaspar et al. 1984 Table 1
Soybean	Field	Silt loam	2.0–2.5	Kaspar et al. 1984, Table 4
Soybean	Plastic tubes, growth chamber	Perlite/clay loam	2.4–3.6	Stone and Buttery, 1986, Table 3
Soybean	Plastic tubes, field	Perlite/clay loam	1.8–2.7	Stone and Buttery, 1986, Table 3
Soybean	Plastic tubes, field	Vermiculite/ perlite	0.4–2.4	Buttery and Stone, 1988, Table 2
Soybean	Rhizotron (July 14–28)	Silt loam	3.7–5.3	Kaspar et al. 1978, Table 1
Soybean	Field	Clay loam	2.5–5.0	Mitchell and Russell, 1971
Pea (*Pisum sativum*)	Root boxes	Solution	2.4	Gerard, 1971
Lettuce (*Lactuca sativa* L.)	Glass tubes	Sandy loam	0.8–2	Carr and Dodds, 1983
Cotton (*Gossypium hirsutum* L.)	Root boxes	Solution	7.9	Gerard, 1971
Tomato (*Lycopersicon esculentum* L.)	Rhizotron	Loamy sand	8.6	Taylor et al. 1970

dominance generally wanes with depth. Some dicots develop adventitious root systems on stolons, runners, or the hypocotyl. The ability of horticulturalists to "root" cuttings is based on this trait.

Rates of taproot elongation are generally of the order of several centimeters day^{-1} (Table 1). Factors such as high soil strength and low temperature cause elongation rates to be generally lower in the field than in the laboratory.

Mitchell and Russell (1971) gave an excellent description of the early development of the soybean taproot at 21 days after planting. The radicle was strongly geotropic and grew vertically downward at 2.5 to 5.0 cm day^{-1}. Several days after germination, lateral roots appeared on the upper 10 to 15 cm of the taproot in four vertical rows at 90-degree angles around the taproot circumference. These branch roots grew laterally for 35 to 40 cm, turned downward, and then grew vertically.

Taproot penetration slowed after about 60 days of growth and these strong laterals eventually assumed the role of depth penetration.

Dicot roots generally undergo substantial secondary growth which provides for added layers of vascular tissue to support new shoot tissues. The stele of a dicot primary root consists of a central xylem which is somewhat star-shaped with patches of phloem between the arms of the star. Between the xylem and phloem, a layer of meristematic tissue, the cambium, develops and divides off new xylem cells to the inside and new phloem cells to the outside. Eventually this radial growth causes the old exterior primary tissues to be sloughed off and a woody root with an external bark (phloem) results. This root can grow quite large in trees and perennial shrubs.

Fleshy taproots undergo specialized secondary growth that can be illustrated by sugarbeet (*Beta vulgaris* L.), as described by Rapaport and Loomis (1986, 1987). This taproot contains concentric rings of cambia which form centrifugally. Thus, the beet root consists of concentric rings of vascular tissue separated by broad bands of large-celled storage parenchyma that enlarge to provide storage cells. Most fleshy taprooted crops are biennials, i.e., they grow vegetatively one year while they store reserves in below-ground roots and then they flower the second year.

B. Lateral Branching

Lateral roots are produced from primordia that usually originate in the root pericycle at the points of the xylem "star" which contain protoxylem elements. This orientation leads to the four-ranked pattern observed in soybean roots by Mitchell and Russell (1971). A pattern is often found in the spacing of these laterals up and down the root. They are arranged in groups with each group consisting of one lateral from each protoxylem pole. For example, if the "star" has four arms, there would be four laterals near one another on the axis, followed by a space with no roots, followed by another clump of four laterals. This sort of pattern, well-illustrated in Mallory et al. (1970, Fig. 7), may not occur in some species or conditions and may be obscured by conditions under which some of the root primordia remain dormant. These patterns may be under control of such hormones as auxin and cytokinin (Hinchee and Rost, 1986).

Lateral root primordia are produced behind the region of root hairs and require a set amount of developmental time to appear and elongate to produce laterals. For example, MacLeod and Thompson (1979) found that *Vicia faba* began to produce primordia 2.6 cm behind the apex on the fifth day after germination and that these primordia finally appeared 7.7 cm behind the apex seven days after germination. Generally, in this study, about three days elapsed between primordium formation and appearance of the resulting lateral root in three legume species. Klasova et al. (1972) found that two days were required for *Pisum sativum* L.

Differences between these studies could have resulted from differences in the temperatures or cultivars used in each study.

Generally, lateral roots are produced and grow at right angles to the parent root. However, the laterals produced on the taproots of dicots generally grow horizontally for a period and then turn downward. For example, Mitchell and Russell (1971) found that the lateral roots of the soybean taproot grew horizontally for 40 to 50 days before turning downward about 35 to 40 cm away from the taproot. This pattern provides a structure similar to the axes of monocots in that the laterals explore a cylinder of soil outside of the one explored by the taproot. Kaspar et al. (1981) showed that these lateral soybean roots actually grow at an angle to the horizontal and that the angle depends on temperature.

C. Anatomical Features

The principal features which distinguish dicot from monocot roots include the vascular arrangement with its star-shaped xylem in the stele and the occurrence of significant secondary growth. Endodermal features of dicots are similar to monocots (Scott and Peterson, 1979), but little detailed research has been done on dicot endodermal development. The development of vascular connections between the xylem elements of the parent and branch roots has been described for tomato (*Lycopusicon esculentum* Mill.) (Byrne et al., 1982) and soybean (Byrne et al., 1977) and involves events similar to those described for monocots.

Little research has been done on the secondary growth in roots after the initial phases of the formation of a continuous cambium between the xylem and phloem tissues. Generally, woody roots have a structure similar to woody stems with a central woody xylem surrounded by bark (Rapoport and Travis, 1984). One important feature of woody roots is that the secondary vascular tissues of the lateral roots are laid down in a smooth continuous elbow in the upward direction but not in the downward direction so that there is less resistance to water flow in the upward than in the downward direction (Klepper, 1983).

D. Number of Roots

The branching frequency (number mm^{-1}) of lateral roots varies with species and growing conditions. MacLeod and Thomson (1979) found values ranging from 39 to 68 mm^{-1} for three legume species. Geisler and Krützfeldt (1983) found 25 to 39 mm^{-1} for field beans (*Vicia faba* L.) and Blakely et al. (1982) found about 60 mm^{-1} for radish seedlings. High fertility increases the number of laterals in field beans (Geisler and Krützfeldt, 1983) and in tomatoes (1978); both cultivar and soil texture can affect lineal frequency (Gulmon and Turner, 1978). In a study of forage legumes, Pinkerton and Simpson (1981) found that some species responded to liming of acid subsoils by increasing the fineness of the root system. Branching frequency also depends on temperature, with a maximum at about 35°C of about 15 mm^{-1} for 10-day-old seedlings of guar (*Cyamopsis tetragonoloba* (L). Taub.) (Stafford and McMichael, 1990) and at 20°C (the highest temperature tested) for *Pinus radiata* 55 days after transplanting (Nambiar et al., 1979). Generally, impacts of soil type and physical condition are greater than varietal impacts on root patterns (McIntosh and Miller, 1981).

E. Lengths, Diameters, and Weights of Roots

Elongation rates, diameters, and dry weights of dicot roots depend on plant vigor, shoot environment, and local soil conditions. Because of secondary growth, taproots and first-order laterals are generally much thicker and heavier than higher-order laterals, especially in the upper part of the soil profile, but they comprise a smaller part of the absorbing root length. For example, Fiscus (1981) found that about two-thirds of the total root area of *Phaseolus vulgaris* L. was made up of roots with an average diameter of about 0.2 mm, about one-fifth was roots with a diameter of 0.5 mm, and the rest (less than 15%) was made up of larger roots. Similar information has been reported for potato (*Solanum tuberosum* L.) roots where 91% of the roots had diameters smaller than 0.44 mm, with the most frequent class between 0.12 and 0.20 mm, and only about 1% of the diameters exceeded 1.3 mm (Vos and Groenwold, 1986).

Lineal densities of dicot roots are generally higher than values for monocot roots. For example, Vos and Groenwold (1986, Table 4) found values ranging from 5.5 to 12.7 $mg\,m^{-1}$ for potato under field conditions, with the higher values being found at the depth where one would expect to find the main taproot and laterals. In a pot experiment, Borkert and Barber (1985) found values ranging from 1.9 to 4.4 $mg\,m^{-1}$ for soybean, with generally lower values (more fine roots) under high phosphorus nutrition. Average root radii in this pot experiment ranged from 0.4 to 0.8 mm for the low and high values of lineal density respectively.

F. Dry Matter Partitioning

Dicots show the same general trend of a decreasing root to shoot ratio as the season progresses that monocots do. For example, in Syria, spring-sown chickpeas (*Cicer arietinum*) had root to shoot ratios of 0.59 in mid-April, 0.41 in May, and 0.18 at maturity (Brown et al., 1989). For field-grown soybean plants, ratios dropped from 0.26 in the V6 stage to 0.11 in the V15R2 stage (Sivakumar et al., 1977). At harvest the average root to shoot ratio for six cowpea (*Vigna unguiculata* Walp.) cultivars was 0.026 (Kahn and Stoffella, 1987). Borkert and Barber (1985) showed that this ratio decreased from 0.21 for no added phosphorus to 0.10 for

960 mg kg^{-1} added in pot studies on 24-day-old soybean plants. With added nitrogen, Kuchenbuch et al. (1988) found that the root to shoot ratio also decreased in the sunflower (*Helianthus annuus* L.). With elevated atmospheric CO_2 concentrations, Idso et al. (1988) found that woody-rooted dicots such as cotton and soybean maintained the same root to shoot ratios at 640 mL L^{-1} CO_2 as at 340 mL L^{-1}, while fleshy taproots such as radish and carrot had ratios with the higher CO_2 concentration, about one-third higher than with normal CO_2 levels. Huber (1983) has suggested that a relationship may exist between the starch/sucrose partitioning in leaves and the root to shoot partitioning of dry matter. Species with high levels of leaf starch generally have low relative root weights, possibly because leaf starch is used at night for shoot growth in preference to root growth.

G. Modeling Concepts for Dicot Root Systems

We know enough about a few dicots such as soybean or cotton to make mathematical simulations of root growth. Generally, the dicot root system consists of vertically-growing taproots or taproots and main laterals which provide increases in rooting depth. Then, higher-order laterals provide increases in root length density at a given depth. Presumably, the time from appearance of a parent root segment to the first appearance of laterals on that segment can be calculated using GDD or some other measure of developmental time. Detailed account must be taken of senescence of lateral roots as related to time and plant stress factors (Huck et al., 1987). The first detailed root model, RHIZOS, was written to simulate cotton root growth (Bar-Yosef et al., 1982), and it provides for average root length densities of the crop in a two-dimensional array across the crop row. Huch and Hillel (1983) also developed a root growth model for cotton where carbohydrates were partitioned to roots by taking into account root to shoot water potential relationships. A further modification of these concepts was provided in ROOTSIMU V4.0 (Hoogenboom and Huck, 1986) and ROOTSIMU V4.3 (Hoogenboom et al., 1987) which account for localized restrictions in root extension as the soil dries. A more complete account of root modelling can be obtained from recent reviews (Klepper and Rickman, 1990, 1991).

IV. Needed Research

A. Root System Developmental Patterns and Field Soil Environments

Our understanding of the soil physical and chemical factors and genetic plant factors that govern rooting patterns of crops is fairly adequate (Hamblin, 1985; Klepper, 1990). From greenhouse, growth chamber, laboratory, and field studies, we have found that plant root growth is especially responsive to soil strength, water content, and temperature under most agronomic situations, but can be drastically affected by other conditions such as residual chemicals, acidity, flooding (anaerobiosis), and numerous other factors (Klepper, 1990).

Although the impacts of individual factors may have been well characterized for particular soils, we need to develop generalizations that would apply over a range of soil types and conditions for modeling purposes (Bar-Yosef and Lambert, 1981). We need predictors of soil physical properties such as strength, hydraulic conductivity, and water retention characteristics on the basis of soil texture and other information readily available from soil databases. Examples of such approaches include work on predicting bulk density changes from fundamental parameters (Jones, 1983; Gupta and Allmaras, 1987; Smith, 1987). Also, more physically-based concepts need to be applied to the interactions between growing roots and soil, as illustrated by the work of Dexter and Hewitt (1978), Richards and Greacen (1986), and Rickman et al. (1991). Availability of these relationships will allow dynamic root growth and function models to be written.

Greater understanding is also needed of soil structure, especially of the larger pores left by previous biological activity or of soil cracks large enough to allow root entry (Edwards and Lofty, 1978). These pores can be very important channels of low resistance for root elongation (Ehlers et al., 1983) and may control the spatial distribution of roots and their functions (Geisler and Daroussis, 1981; Hamblin, 1985; Passioura, 1985).

B. Root Exudates and Rhizosphere Dynamics

Some material is lost from living roots both during and after extension growth. These materials range from intact cells sloughed or scraped from the root surfaces, to globular or fibrillar mucilages on root surfaces, to soluble materials detected in the rhizosphere. The rhizosphere can be defined as the cylinder of soil adjacent to the root and affected by root activity. Exuded materials consist of such compounds as sugars and amino acids. These compounds provide a food base for bacteria and fungi that reside in the rhizosphere (Helal and Sauerbeck, 1986).

The exudates of some crop species have been chemically characterized (e.g., see Matsumoto et al., 1979; Rougier and Chabond, 1985) and studies have shown that quantities of these exudates depend on species, root age, and environmental conditions (Vancura et al., 1977; Keith and Oades, 1986; Biondini et al., 1988). Exudates can constitute up to one-quarter of net photosynthesis (Barber and Martin, 1976; Merckx et al., 1986) and thus can represent a significant proportion of available photosynthate (Milchunas et al., 1985; Keith and Oades, 1986).

In spite of excellent studies on exudates, more research is needed on the role which sloughed cells, exudates, and mucilages play in the normal functioning of the roots to elucidate the role of these materials in the rhizospheres of field-grown crops. Do exuded chelates play a significant role in nutrient uptake (e.g., Jauregui and Reisenauer, 1982), and is the composition of the exudates controlled by plant or soil factors? What are the functions, if any, of the exudates in protecting roots from microbial attack or in increasing nodulation or mycorrhizal associations? How do the exudates of monocots and dicots differ in composition and physiology? How important are genetic factors in defining the suites and amounts of exudates in the rhizosphere and what sorts of selections could be of advantage to agriculture? In short, we need to understand enough about root exudates to be able to manipulate them to optimize the rhizosphere for crop production and/or for the enhancement of desirable soil properties.

C. Root Senescence Patterns

Little quantitative work has been done on root death through natural senescence, although we do know that the loss of the cortex of monocot roots is a natural process (Henry and Deacon, 1981) that is accelerated by low local soil fertility (Gillespie and Deacon, 1988). Cortical senescence may be related to the development of the endodermis which reduces the flux of food materials from the vascular tissues to the outer root tissues.

Death of entire roots also occurs in both monocots and dicots, but we do not know what determines the fact that certain lateral roots are ephemeral and others are long-standing. In a study of two C3 cool-season perennial grasses, Distel and Fernandez (1988) demonstrated that root decay rates ($mm\,d^{-1}$) in the top 0.6 m of the profile were roughly equal to rates of root growth when averaged over a two-year period so that, by dynamic equilibrium, the amount of living root material was relatively constant over the two-year period.

Root senescence, whether of cortical tissues or of entire laterals or axes, adds significant quantities of organic material to soils and provides a food base for soil biota. Relatively little work has been done to quantify the role of root death in nutrient and organic-matter cycling in cropped and native soils, especially when contrasted to the considerable research on the decay of shoot residues.

D. Root Composition and Decay Dynamics

We measure dry matter distribution to the roots and changes in root length density with depth over time in the growing season and rarely stop to think about the role which these belowground organic materials may play in the long-term changes in soil properties. It is probably not a coincidence that our richest, darkest prairie soils were produced beneath grass-dominated vegetations. Grasses have half or more of their biomass belowground (Schulze, 1983) and root materials have been found to contribute more to soil organic matter than shoots (Barber, 1979; Buyanovsky and Wagner, 1987). Furthermore, grasslands are subject to dry periods when the decay of residues is retarded and to cold periods when low temperatures slow decay processes.

Plants produce considerable biomass in the top half-meter of soil with decreasing amounts below that depth. In the native prairie species, there were deep perennial roots at all times in the season. In the natural ecosystem, high rates of decay and rapid shoot and root growth were coincidental with wet periods. Nutrients held in old root material were released by decay activities at about the same time that the annual and perennial prairie grasses were undergoing growth spurts, thus minimizing the potential for leaching losses from the root zone. This natural cycle has been broken with annual cropping because there is a considerable period when wet fields are either bare or have only shallow-rooted seedlings to utilize the nutrients released from the "root-bank." Usually there is a one- or two-year cropping cycle with a period of root growth and soil exploration coinciding with the cropping season. This cycle is followed by a period of root death and decay with no living roots present to extract the nutrients released, in contrast to the continual dynamic root turnover found in the prairie. Since nitrogen is moved out of the roots during grain filling (Smith et al., 1983), the C:N ratios of root material at the end of the season are higher than they are during the cropping year. We do not know whether nutrients are removed from senescing laterals before they die to be reused elsewhere in the plant, but, by analogy to senescing leaves, one might expect such translocation to occur.

E. Root-Shoot "Signals" and Functional Integration

Root growth and function are both dynamic phenomena (Crossett et al., 1975). When growth of one part of a root system is slowed or prevented, compensatory growth usually occurs elsewhere. For example, in well-watered soil, a vigorous root system is supported in the top part of the soil profile; in a drying profile, that same root system ceases to produce new root material in the top of the profile and accelerates growth of new roots in the bottom of the profile (Klepper, 1990). Similarly, shifts in the root to shoot ratio can be brought about by various environmental shifts such as drought. No one has unequivocally established the mechanism by which these shifts occur. One concept is that stress decreases the growth of one part and causes an increase in the growth of the remaining parts simply by increasing the carbohydrates and other materials available. Another idea is that hormones such as abscisic acid (ABA) are involved in root to shoot communication (Blackman and Davies, 1985; Zhang and Davies, 1989). Until more definitive research is performed, we cannot differentiate these two hypotheses from one another. It is well known that such growth substances as ABA and indoleacetic acid (IAA) can regulate root elongation (Saugy et al., 1989). These hormones are also involved in the root growth response to gravity (Pilet and Barlow, 1987), along with other factors such as calcium (Leopold and Wettlaufer, 1988).

With respect to root function, hardly any research has examined the integration of the entire system except in the area of water uptake. Extensive research has shown that water uptake into plant roots occurs primarily in response to water potential gradients between the bulk soil and the root interior. Plant activities such as transpiration and/or production of osmotically-active materials cause the water potential in the root xylem to be low enough to cause water to move into the roots from relatively moist soil. Since the xylem is a more or less continuous water column from the root to the shoot, "communication" via xylem tensions (negative pressures) could provide rapid feedback between the shoot and various parts of the root system. Also, uptake in the upper part of the profile automatically reduces the xylem potential in the lower part of the root system and thus reduces uptake in the lower part of the soil profile.

Mechanisms governing integration of nutrient uptake by the roots and the nutritional balances between the roots and shoots are less clear. If one part of a root system is in a part of the profile with a high nitrogen content, how is the total uptake of nitrogen determined? What mechanism governs the fractional uptake from high- and low-nitrogen parts of the system? Research in this area is needed for improved precision in modeling and in configuring fertilizer applications for optimal root uptake.

F. Root System Management

Roots must become a more prominent part of the thinking of agricultural managers so that we can (1) design and install desirable root traits into varieties by use of genetic engineering (O'Toole and Bland, 1987); (2) manipulate root system distributions in the soil profile to optimize the abilities of roots to take up water and nutrients; and (3) understand the role of decaying root and root materials in the determination of such soil properties as organic matter composition and native fertility. We are quite accustomed to the idea of designing crop canopies for the ease of harvest or for improved canopy capture of sunlight. We need to apply our plant breeding capabilities to the root system, but only after we have a better understanding of the desirability of the traits that we can manipulate.

References

Anderson-Taylor, G., and C. Marshall. 1983. Root-tiller interrelationships in spring barley (*Hordeum distichum* (L.) Lam.). *Ann. Bot.* 51:47–58.

Barber, D.A., and J.K. Martin. 1976. The release of organic substances by cereal roots into soil. *New Phytol.* 76:69–80.

Barber, S.A. 1979. Corn residue management and soil organic matter. *Agron. J.* 71:625–627.

Barraclough, P.B. 1984. The growth and activity of winter wheat roots in the field: Root growth of high-yielding crops in relation to shoot growth. *J. Agric. Sci. Camb.* 103:419–422.

Barraclough, P.B., and R.A. Leigh. 1984. The growth and activity of winter wheat roots in the field: The effect of sowing date and soil type on root growth of high-yielding crops. *J. Agric. Sci.* 103:59–74.

Bar-Yosef, B., and J.R. Lambert. 1981. Corn and cotton root growth in response to soil impedance and water potential. *Soil Sci. Soc. Am. J.* 45:930–935.

Bar-Yosef, B., J.R. Lambert, and D.N. Baker. 1982. Rhizos: A simulation of root growth and soil processes. Sensitivity analysis and validation for cotton. *Trans. ASAE* 25:1268–1273, 1281.

Belford, R.K., B. Klepper, and R.W. Rickman. 1987. Studies of intact shoot-root systems of field-grown winter wheat. II. Root and shoot developmental patterns as related to nitrogen fertilizer. *Agron. J.* 79:310–319.

Biondini, M., D.A. Klein, and E.F. Redente. 1988. Carbon and nitrogen losses through root exudation by *Agropyron Cristatum, A. Smithii*, and *Bouteloua Gracilis*. *Soil Biol. Biochem.* 20:477–482.

Blackman, P.G., and W.J. Davies. 1985. Root to shoot communication in maize plants of the effects of soil drying. *J. Exp. Bot.* 36:39–48.

Blakely, L.M., M. Durham, T.A. Evans, and R.M. Blakely. 1982. Experimental studies on lateral root formation in radish seedling roots. I. General methods, developmental stages, and spontaneous formation of laterals. *Bot. Gaz.* 143:341–352.

Borkert, C.M., and S.A. Barber. 1985. Soybean shoot and root growth and phosphorus concentration as affected by phosphorus placement. *Soil Sci. Soc. Am. J.* 49:152–155.

Brown, S.C., P.J. Gregory, P.J.M. Cooper, and J.D.H. Keatinge. 1989. Root and shoot growth and water use of chickpea (*Cicer arietinum*) grown in dryland conditions: Effects of sowing date and genotype. *J. Agric. Sci.* 113:41–49.

Buttery, B.R., and J.A. Stone. 1988. Some effects of nitrate and light intensity on soybean root growth and development. *Plant Soil* 106:291–294.

Buyanovsky, G.A., and G.H. Wagner. 1987. Carbon transfer in a winter wheat (*Triticum aestivum*) ecosystem. *Biol. Fert. Soils* 5:76–82.

Byrne, J.M., J.M. Byrne, and D.P. Emmitt. 1982. Development and structure of the vascular connection between the primary and lateral root of *Lycopersicon esculentum*. *Am. J. Bot.* 69:287–297.

Byrne, J.M., T.C. Pesacreta, and J.A. Fox. 1977. Development and structure of the vascular connection between the primary and secondary root of *Glycine max* (L.) Merr. *Am. J. Bot.* 64:946–959.

Caradus, J.R. 1981. Root growth of white clover (*Trifolium repens* L.) lines in glass-fronted containers. *New Zeal. J. Agr. Res.* 24:43–54.

Carr, M.K.V., and S.M. Doods. 1983. Some effects of soil compaction on root growth and water use of lettuce. *Exp. Agric.* 19:117–130.

Christie, E.K. 1975. Physiological responses of semiarid grasses. II. The pattern of root growth in relation to external phosphorus concentration. *Aust. J. Agric. Res.* 26:437–446.

Clark, L.H., and W.H. Harris. 1981. Observations on the root anatomy of rice (*Oryza sativa* L.). *Am. J. Bot.* 68:154–161.

Clarkson, D.T., and J. Sanderson. 1974. The endodermis and its development in barley roots as related to radial migration of ions and water. *In* J. Kolek (Ed.). *Structure and Function of Primary Root Tissues*. Veda, Publishing House of the Slovark Academy of Sciences, Bratislavia, pp. 87–100.

Cornish, P.S., J.R. McWilliam, and H.B. So. 1984. Root morphology, water uptake, growth, and survival of seedlings of ryegrass and phalaris. *Aust. J. Agric. Res.* 35:479–492.

Crossett, R.N., D.J. Campbell, and H.E. Stewart. 1975. Compensatory growth in cereal root systems. *Plant Soil.* 42:673–683.

Dexter, A.R., and J.S. Hewitt. 1978. The deflection of plant roots. *J. Agric. Eng. Res.* 23:17–22.

Distel, R.A., and O.A. Fernández. 1988. Dynamics of root growth and decay in two grasses native to semi-arid Argentina. *Aust. J. Ecol.* 13:327–336.

Drew, M.D., and L.R. Saker. 1975. Nutrient supply and the growth of the seminal root system in barley. II. Localized compensatory increases in lateral root growth and rates of nutrient uptake when nitrate supply is restricted to only part of the root system. *J. Exp. Bot.* 26:79–90.

Edwards, C.A., and J.R. Lofty. 1978. The influence of arthropods and earthworms upon root growth of direct drilled cereals. *J. Appl. Ecol.* 15:789–795.

Ehlers, W., U. Kopke, F. Hesse, and W. Bohm. 1983. Penetration resistance and root growth of oats in tilled and untilled loess soil. *Soil Till. Res.* 3:261–275.

Ennik, G.C., and T.B. Hofman. 1983. Variation in the root mass of ryegrass types and its ecological consequences. *Neth. J. Agric. Sci.* 31:325–334.

Feil, B., and G. Geisler. 1988. Root growth of seedlings of old and new winter wheat cultivars and a spelt wheat at varying levels of nitrogen. *J. Agron. Crop Sci.* 161:264–272.

Fiscus, E.L. 1981. Analysis of the components of area growth of bean root systems. *Crop Sci.* 21:909–914.

Fritsch, R. 1977. Root anatomy of *Triticum* L. and *Aegilops* L. (*Gramineae*). *Flora Bd.* 166:289–326.

Geisler, G., and N. Daroussis. 1981. Influence of pore size distribution, water content, and air filled pore volume on morphological characters of the root system and growth of maize, broad beans, and barley. *J. Agron. Crop Sci.* 150:457–473.

Geisler, G., and B. Krutzfeldt. 1983. Investigations into the effect of "nitrogen" on the morphology, dry matter formation, and nutrient uptake efficiency of the root systems of maize, spring barley, and field bean varieties, having regard to temperature conditions. I. Root morphology. *J. Agron. Crop Sci.* 152:336–353.

Geisler, G., and B. Krutzfeldt. 1984. Wirkungen von Stickstoff auf die Morphologie und die Trockenmassebildung der Wurzelsysteme von Mais-, Sommergersten- und Ackerbohnen-Sorten unter Berucksichtigung der Temperatur: 2. Trockenmassebildung. *J. Agron. Crop Sci.* 153:90–104.

Gerard, C.J. 1971. Influence of osmotic potential, temperature, and calcium on growth of plant roots. *Agron. J.* 63:555–558.

Gerwitz, A., and E.R. Page. 1974. An empirical mathematical model to describe plant root systems. *J. Appl. Ecol.* 11:773–781.

Gillespie, I.M.M., and J.W. Deacon. 1988. Effects of mineral nutrients on senescence of the cortex of wheat roots and root pieces. *Soil Biol. Biochem.* 20:525–531.

Girardin, P., B. Morel-Fourrier, M.-O. Jordan, and B. Millet. 1987. Adventitious root development in *Zea mays*. *Agronomie* 7:353–360.

Gregory, P.J. 1983. Response to temperature in a stand of pearl millet (*Pennisetum typhoides* S. & H.). III. Root development. *J. Exp. Bot.* 34:744–756.

Gregory, P.J. 1986. Response to temperature in a stand of pearl millett (*Pennisetum typhoides* S. & H.). VIII. Root growth. *J. Exp. Bot.* 37:379–388.

Gulmon, S.L., and N.C. Turner. 1978. Differences in root and shoot development of tomato (*Lycopersicon esculentum* L.) varieties across contrasting soil environments. *Plant Soil* 49:127–136.

Gupta, S., and R.R. Allmaras. 1987. Models to assess the susceptibility of soils to excessive compaction. *Adv. Soil Sci.* 6:65–100.

Hackett, C. 1972. A method of applying nutrients locally to roots under controlled conditions, and some morphological effects of locally applied nitrate on the branching of wheat roots. *Aust. J. Biol. Sci.* 25:1169–1180.

Hackett, C., and B.O. Bartlett. 1971. A study of the root system of barley III. Branching pattern. *New Phytol.* 70:409–413.

Hamblin, A.P. 1985. The influence of soil structure on water movement, crop root growth, and water uptake. *Adv. Agron.* 38:95–158.

Helal, H.M., and D. Sauerbeck. 1986. Effect of plant roots on carbon metabolism of soil microbial biomass. *Z. Pflanzenernaehr. Bodenk.* 149:181–188.

Henry, C.M., and J.W. Deacon. 1981. Natural (non-pathogenic) death of the cortex of wheat and barley seminal roots, as evidenced by nuclear staining with acridine orange. *Plant Soil* 60:255–274.

Hinchee, M.A.W., and T.L. Rost. 1986. The control of lateral root development in cultured pea seedlings. I. The role of seedling organs and plant growth regulators. *Bot. Gaz.* 174:137–147.

Hoogenboom, G., and M.G. Huck. 1986. ROOTSIMU V4.0 a dynamic simulation of root growth, water uptake, and biomass partitioning in a soil-plant-atmosphere continuum: Update and documentation. *Agronomy and Soils Dept. Ser. No. 109*, Alabama Agricultural Experiment Station, Auburn.

Hoogenboom, G., M.G. Huck, and C.M. Peterson. 1987. Root growth rate of soybean as affected by drought stress. *Agron. J.* 79:607–614.

Huber, S.C. 1983. Relation between photosynthetic starch formation and dry-weight partitioning between the shoot and root. *Can. J. Bot.* 61:2709–2716.

Huck, M.G., and D. Hillel. 1983. A model of root growth and water uptake accounting for photosynthesis, respiration, transpiration, and soil hydraulics. *Adv. Irrig.* 2:273.

Huck, M.G., G. Hoogenboom, and C.M. Peterson. 1987. Soybean root senescence under drought stress. *In Minirhizotron Observation Tubes: Methods and Applications for Measuring Rhizosphere Dynamics*. ASA Special Publication No. 50, pp. 109–121.

Idso, S.B., B.A. Kimball, and J.R. Maurey. 1988. Effects of atmospheric CO_2 enrichment on root:shoot ratios of carrot, radish, cotton, and soybean. *Agric. Ecosystems Environ.* 21:293–299.

Jauregui, M.A., and H.M. Reisenauer. 1982. Dissolution of oxides of manganese and iron by root exudate components. *Soil Sci. Soc. Am. J.* 46:314–317.

Johnson, I.R., and J.H.M. Thornley. 1983. *Plant Cell Environ.* 6:721–729.

Jones, C.A. 1983. Effect of soil texture on critical bulk densities for root growth. *Soil Sci. Soc. Am. J.* 47:1208–1211.

Jordan, M.-O. 1987. Growth and development of the corn root system. I. Importance of some parameters concerning the plant. *Agronomie* 7:365–371.

Jordan, W.R., M. McCrary, and F.R. Miller. 1979. Compensatory growth in the crown root system of sorghum. *Agron. J.* 71:803–806.

Jupp, A.P., and E.I. Newman. 1987. Morphological and anatomical effects of severe drought on the roots of *Lolium perenne* L. *New Phytol.* 105:393–402.

Kahn, B.A., and P.J. Stoffella. 1987. Root morphological characteristics of field-grown cowpeas. *J. Am. Soc. Hort. Sci.* 112:402–406.

Kaigama, B.K., I.D. Teare, L.R. Stone, and W.L. Powers. 1977. Root and top growth of irrigated and nonirrigated grain sorghum. *J. Crop Sci.* 17:555–559.

Kaspar, T.C., C.D. Stanley, and H.M. Taylor. 1978. Soybean root growth during the reproductive stages of development. *Agron. J.* 70:1105–1107.

Kaspar, T.C., H.M. Taylor, and R.M. Shibles. 1984. Taproot-elongation rates of soybean cultivars in the glasshouse and their relation to field rooting depth. *Crop Sci.* 24:916–920.

Kaspar, T.C., D.G. Woolley, and H.M. Taylor. 1981. Temperature effect on the inclination of lateral roots of soybeans. *Agron. J.* 73:383–386.

Kawata, S., J. Harada, and K. Yamazaki. 1978. On the number and the diameter of crown root primordia in rice plants. *Japan. J Crop Sci.* 47:644–654.

Kawata, S., O. Sasaki, and K. Yamazaki. 1977. On the structure of the crown root and the lateral root, and the vessel connection between them, in rice plants. *Japan. J. Crop Sci.* 46:569–579.

Keith, H., and J.M. Oades. 1986. Input of carbon to soil from wheat plants. *Soil Biol. Biochem.* 18:445–449.

Klasova, A., J. Kolek, and J. Klas. 1972. Time dynamics of primary root branching in *Pisum sativum* L. *Biologia Plantarum (Praha)* 14:249–253.

Klepper, B. 1983. Managing root systems for efficient water use: Axial resistances to flow in root systems—anatomical considerations. *In Limitations to Efficient Water Use in Crop Production.* ASA, CSSA, and SSSA, Madison, Wisconsin, pp. 115–125.

Klepper, B. 1990. Root growth and water uptake. *In* B.A. Stewart and D.R. Nielsen (Eds.). *Irrigation of Agricultural Lands.* Am. Soc. Agron., Madison, Wisconsin, pp. 281–322.

Klepper, B. 1991. Root-shoot relationships. *In* Y. Waisel, U. Kafkafi, and A. Eshel (Eds.). *Plant Roots: The Hidden Half.* Marcel Dekker, New York, pp. 265–286.

Klepper, B., and R.W. Rickman. 1990. Modeling root growth and function. *Adv. Agron.* 44:113–131.

Klepper, B., and R.W. Rickman. 1991. Predicting root development of crop plants. *In* Tom Hodges (Ed.). *Physiological Aspects of Predicting Crop Phenology.* CRC Press, Boca Raton, FL, pp. 85–99.

Klepper, B., R.K. Belford, and R.W. Rickman. 1984. Root and shoot development in winter wheat. *Agron. J.* 76:117–122.

Kuchenbuch, R., W. Weigelt, and J. Jung. 1988. Modification of root-shoot-ratio of sunflower (*Helianthus annuus*, L.) by nitrogen supply and a triazole-type plant growth regulator. *Z. Pflanzenernahr. Bodenk.* 151:391–394.

Leopold, A.C., and S.H. Wettlaufer. 1988. Diagravitropism in corn roots. *Plant Physiol.* 87:803–805.

Macduff, J.H., A. Wild, M.J. Hopper, and M.S. Dhanoa. 1986. Effects of temperature on parameters of root growth relevant to nutrient uptake: Measurements on oilseed rape and barley grown in flowing nutrient solution. *Plant Soil* 94:321–332.

MacLeod, R.D., and A. Thompson. 1979. Development of lateral root primordia in *Vicia faba, Pisum sativum, Zea mays*, and *Phaseolus vulgaris*: Rates of primordium formation and cell doubling times. *Ann. Bot.* 44:435–449.

Mallory, T.E., S.-H. Chiang, E.G. Cutter, and E.M. Gifford, Jr. 1970. Sequence and pattern of lateral root formation in five selected species. *Am. J. Bot.* 57: 800–809.

Matsumoto, H., K. Okada, and E. Takahashi. 1979. Excretion products of maize roots from seedling to seed development stage. *Plant Soil* 53:17–26.

May, L.H., F.H. Chapman, and D. Aspinall. 1965. Quantitative studies of root development. I. The influence of nutrient concentration. *Aust. J. Biol. Sci.* 18:25–35.

McIntosh, M.S., and D.A. Miller. 1980. Development of root-branching in three alfalfa cultivars. *Crop Sci.* 20:807–809.

McIntosh, M.S., and D.A. Miller. 1981. Genetic and soil moisture effects on the branching-root trait in alfalfa. *Crop Sci.* 21:15–18.

Merckx, R., J.H. van Ginkel, J. Sinnaeve, and A. Cremers. 1986. Plant-induced changes in the rhizosphere of maize and wheat. I. Production and turnover of root-derived material in the rhizosphere of maize and wheat. *Plant Soil* 96: 85–93.

Milchunas, D.G., W.K. Lauenroth, J.S. Singh, C.V. Cole, and H.W. Hunt. 1985. Root turnover and production by ^{14}C dilution: Implications of carbon partitioning in plants. *Plant Soil* 88:353–365.

Mirhadi, M.J., and Y. Kobayashi. 1980. The relationship between the growth of different plant organs of grain sorghum hybrid H-726. *Japan. J. Crop Sci.* 49: 420–427.

Mitchell, R.L., and W.J. Russell. 1971. Root development and rooting patterns of soybean (*Glycine max* (L.) Merrill) evaluated under field conditions. *Agron. J.* 63:313–319.

Mosher, P.N., and M.H. Miller. 1972. Influence of soil temperature on the geotropic response of corn roots (*Zea mays* L.). *Agron. J.* 64:459–462.

Nambiar, E.K.S., G.D. Bowen, and R. Sands. 1979. Root regeneration and plant water status of *Pinus radiata* D. Don seedlings transplanted to different soil temperatures. *J. Exp. Bot.* 30:1119–1131.

Newman, P.R., and L.E. Moser. 1988. Seedling root development and morphology of cool-season and warm-season forage grasses. *Crop Sci.* 28: 148–151.

O'Toole, J.C., and W.L. Bland. 1987. Genotypic variation in crop plant root systems. *Adv. Agron.* 41:91–145.

Passioura, J.B. 1985. Roots and water economy of wheat. *In* W. Day and R.K. Atkin (Eds.). *Wheat Growth and Modeling*. Plenum, New York, pp. 185–198.

Picard, D., M.-O. Jordan, and R. Trendel. 1985. Rate of appearance of primary roots of maize. I. Detailed study of one cultivar at one site. Agronomie 5:667–676.

Pilet, P.-E., and P.W. Barlow. 1987. The role of abscisic acid in root growth and gravireaction: A critical review. *Plant Growth Regulation* 6:217–265.

Pinkerton, A., and J.R. Simpson. 1981. Effects of subsoil acidity on the shoot and root growth of some tropical and temperate forage legumes. *Aust. J. Agric. Res.* 32:453–463.

Porter, J.R., B. Klepper, and R.K. Belford. 1986. A model (WHTROOT) which synchronizes root growth and development with shoot development for winter wheat. *Plant Soil* 92:133.

Pulgarin, A., J. Navascues, P.J. Casero, and P.G. Lloret. 1988. Branching pattern in onion adventitious roots. *Am. J. Bot.* 75:425–432.

Rapoport, H.F., and R.S. Loomis. 1986. Structural aspects of root thickening in *Beta vulgaris* L.: Comparative thickening in sugarbeet and chard. *Bot. Gaz.* 147:270–277.

Rapoport, H.F., and R.S. Loomis. 1987. Independence of development in shoot and storage root of *Beta vulgaris*. *Bot. Gaz.* 148:342–345.

Rapoport, H.F., and R.L. Travis. 1984. Alfalfa root growth, cambial activity, and carbohydrate dynamics during the regrowth cycle. *Crop Sci.* 24:899–903.

Richards, B.G., and E.L. Greacen. 1986. Mechanical stresses on an expanding cylindrical root analogue in granular media. *Aust. J. Soil Res.* 24:393–404.

Rickman, R.W., B. Klepper, and R.K. Belford. 1985. Developmental relationships among roots, leaves and tillers in winter wheat. *In* W. Day and R.K. Atkin (Eds.). *Wheat Growth and Modeling*. Plenum Publishing Corporation, New York, pp. 83–98.

Rickman, R.W., S.E. Waldman, and B.L. Klepper. 1992. Calculating daily root length density profiles by applying elastic theory to agricultural soils. *J. Plant Nutr.* (*in press*.)

Robards, A.W., and M.E. Robb. 1972. Uptake and binding of uranyl ions by barley roots. *Science* 178:980–982.

Robards, A.W., S.M. Jackson, D.T. Clarkson, and J. Sanderson. 1973. The structure of barley roots in relation to the transport of ions into the stele. *Protoplasma* 77:291–311.

Rougier, M., and A. Chaboud. 1985. Mucilages secreted by roots and their biological function. *Israel J. Botany* 34:129–146.

Saugy, M., G. Mayor, and P.-E. Pilet. 1989. Endogenous ABA in growing maize roots: light effects. *Plant Physiol.* 89:622–627.

Schulze, E.-D. 1983. Root-shoot interactions and plant life forms. *Neth. J. Agric. Sci.* 4:291–303.

Scott, M.G., and R.L. Peterson. 1979. The root endodermis in *Ranunculus acris*. I. Structure and ontogeny. *Can. J. Bot.* 57:1040–1062.

Sivakumar, M.V.K., H.M. Taylor, and R.H. Shaw. 1977. Top and root relations of field-grown soybeans. *Agron. J.* 69:470–473.

Skiles, J.W., J.D. Hanson, and W.J. Parton. 1982. *In* W.K. Lavenrock, G.V. Skogerboe, and M. Flug (Eds.). *Analysis of Ecological Systems: State of the Art in Ecological Modeling*. Elsevier, Amsterdam, pp. 467–473.

Smith, D.L.O. 1987. Measurement, interpretation and modelling of soil compaction. *Soil Use Manage.* 3:87–93.

Smith, T.L., G.A. Peterson, and D.H. Sander. 1983. Nitrogen distribution in roots and tops of winter wheat. *Agron. J.* 75:1031–1034.

Stafford, R.E., and B.L. McMichael. 1990. Primary root and lateral root development in guar seedlings. *Environ. Exp. Bot.* 30:27–34.

Stone, J.A., and B.R. Buttery. 1986. Some effects of nitrate on soybean root development. *Can. J. Plant Sci.* 66:505–510.

Taylor, H.M., M.G. Huck, B. Klepper, and Z.F. Lund. 1970. Measurement of soil-grown roots in a rhizotron. *Agron. J.* 62:807–809.

Vancura, V., Z. Prikryl, L. Kalachova, and M. Wurst. 1977. Some quantitative aspects of root exudation. *Soil Organisms as Components of Ecosystems. Ecol. Bull. (Stockholm)* 25:381–386.

Vincent, C.D., and P.J. Gregory. 1989a. Effects of temperature on the development and growth of winter wheat roots. I. Controlled glasshouse studies of temperature, nitrogen and irradiance. *Plant Soil* 119:87–97.

Vincent, C.D., and P.J. Gregory. 1989b. Effects of temperature on the development and growth of winter wheat roots. II. Field studies of temperature, nitrogen and irradiance. *Plant Soil* 119:99–110.

Vos, J., and J. Groenwold. 1986. Root growth of potato crops on a marine-clay soil. *Plant Soil* 94:17–33.

Walters, D.R., and P.G. Ayres. 1981. Growth and branching pattern of roots of barley infected with powdery mildew. *Ann. Bot.* 47:159–162.

Yoshida, S., D.P. Bhattacharjee, and G.S. Cabuslay. 1982. Relationship between plant type and root growth in rice. *Soil Sci. Plant Nutri.* 28:473–482.

Zhang, J., and W.J. Davies. 1989. Abscisic acid produced in dehydrating roots may enable the plant to measure the water status of the soil. *Plant Cell Environ.* 12:73–81.

Soil Environment Constraints to Root Growth

Richard W. Zobel

I. Introduction 27
II. The Rhizosphere Environment 29
III. Root System Morphology 36
IV. Root-Soil Interactions 37
V. Conclusions 46
References 48

"Obstacles are what you see when
you take your eyes off the goal."
Anon.

I. Introduction

The concept embodied in the statement "constraints to root growth" is an inherently negative outlook on how biological systems function relative to their surroundings. This concept of one or more constraints is reinforced by a lack of adequate information about how the complex interactions, which take place between plants and their environment, integrate to give the observed responses. For further rapid advances in agricultural productivity and sustainability, with or without reduced economic inputs, a holistic view of the pattern of these interactions will be necessary. Developing such an holistic view will require extensive, wide-interdisciplinary research into the integration of soil characteristics and plant responses. Fortunately, the past extensive research into specific soil physical and chemical properties in relation to crop plant productivity has established a strong framework from which this further needed research can be devel-

Advances in Soil Science, Volume 19

oped. Much of the technology needed to explore root/soil interactions is of recent vintage, with some of the best currently under development.

Historically, plants have produced much lower yields in their natural habitats than when placed in habitats established for their domestication. These new habitats are usually selected to provide reduced competition, improved fertility, and reduced disease incidence, thus allowing increased productivity. The process of domestication, i.e., cultivar selection or breeding, tailors a given plant species into a cultivar(s) which has seed and/or fruit characteristics and shoot morphologies fitted to the needs of the consumer and the available resources (soils, equipment, etc.) of the farmer. As an indirect result of this breeding and selection process, these cultivars would be expected to have improved adaptation to the soil and atmospheric conditions surrounding the fields in which they have been domesticated. One result is that the cultivars become best attuned to the elite soils at experiment stations while becoming (or remaining) more poorly adapted to the average soils found in farmers' fields (Zobel, 1975; Smith and Zobel, 1991).

Most improvements in plant adaptation to edaphic characteristics are accomplished cryptically, i.e., by chance rather than through direct breeder intervention (Zobel, 1975; Smith and Zobel, 1991). More specifically, most changes in edaphic adaptation by crop plants have occurred through selection for yield, plant stature, disease resistance, etc. when the crop is grown on specific soils. Since improved adaptation to existing edaphic conditions enhances these selected characteristics, edaphic adaptation is therefore indirectly selected (cryptically—without knowledge of or planning for such selection). Because the process of selection for adaptation to edaphic characteristics has been, for the most part, indirect, it is probable that plant root systems are, overall, still better adapted to the soils and ecologies of their birth than to the elite soils which are under intense cultivation. During the evolution of existing plant species, a diversity of mechanisms has been adopted by plant communities to adapt their progeny to the encountered range of environmental perturbations. During the breeding for adaptation to agricultural soils, it is likely that only one or several of these mechanisms are cryptically enhanced or suppressed while the majority are selectively neutral relative to the edaphic conditions where selection is taking place (see Wilsie, 1962; Stebbins, 1966, for general discussions of these aspects).

The relative time differential between natural evolution and the forced evolution (breeding) associated with domestication also supports the suggestion that much of the genetic basis for this natural interaction remains intact. Of course, specific exceptions to this assessment exist, such as responsiveness to nitrogenous fertilizers as embodied in the wheat (*Triticum vulgare*) cultivars which spawned the "Green Revolution," specific disease or insect resistance, and changes in nutritional quality.

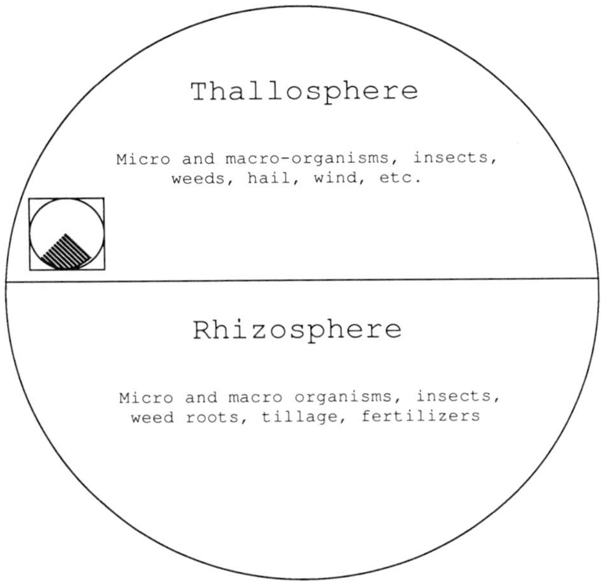

Figure 1. Diagrammatic representation of the biosphere indicating its two main halves. Small inset represents the perceived relationship between the rhizosphere and thallosphere.

These exceptions are always associated with direct breeding efforts that focus on those characteristics. On the whole, the changes man has made in adaptation to edaphic environments while developing the current set of cultivars are probably mostly superficial. If agriculture is to continue to expand and improve, especially onto marginal lands and in a sustainable fashion, this unselected ability to adapt must be accessed. Two general options are available: (1) change the adaptation of cultivars through more directed breeding of root characteristics to meet the needs of problem soils; or (2) change the soil characteristics to more closely meet the needs or adaptation patterns of the plants themselves. In either case, it is critical that we establish the qualitative and quantitative parameters associated with the edaphic characteristics (including levels or quantities) to which given plant species or cultivars are adapted.

II. The Rhizosphere Environment

A. Introduction

Agriculture takes place in the earth's biosphere (Fig. 1). In terms of plant culture, the biosphere can be divided into the rhizosphere and the thallosphere. As intimated in the preceding paragraphs, the preponderance

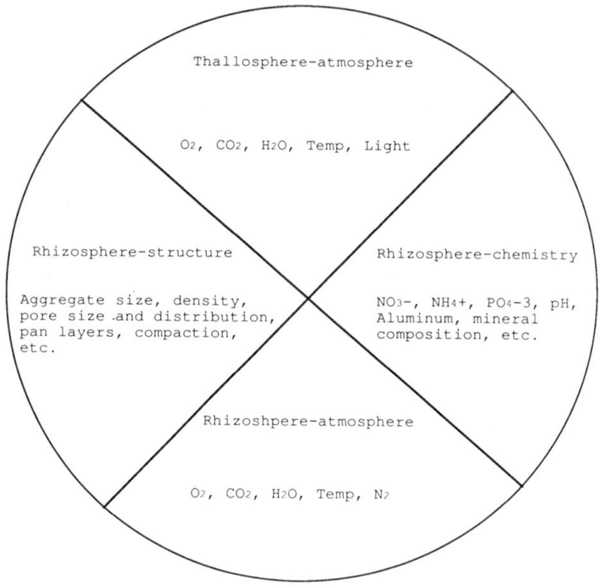

Figure 2. Diagrammatic representation of the main environmental factors which interact with plants to control (modulate) plant growth and development.

of past research has been on shoot characteristics, resulting in the rather biased perception of relative importance represented by the figure in the small inset box of Figure 1, with the shaded area representing the rhizosphere. When direct impact of the environment on the plant is considered, a very different view is obtained (Fig. 2). The primary aspects of the thallosphere environment with which the plant must interact to accomplish growth and development are oxygen, carbon dioxide, water, temperature, and light—in other words, the atmosphere. The sheer complexity of the rhizosphere is augmented by the presence of multitudes of micro- and macro-organisms. However, for conceptual simplicity, no further consideration of interacting organisms and "natural traumas" is made here. The rhizosphere has an atmosphere which is essentially equivalent to that of the thallosphere and, in addition, a physical structure and chemistry with which the plant must also interact (Fig. 2). This makes the complexity of plant/environment interactions greater by far in the rhizosphere than in the thallosphere. Attempts at a holistic view of these interactions usually suffer from the sheer mass of their numbers.

Human intervention in the rhizosphere makes all aspects temporally variable—changing over time. Some characteristics of the rhizosphere environment, however, change so slowly over a cropping season that they appear essentially temporally static (Fig. 3). All soil characteristics tend to be spatially variable on the level of hundreds of square kilometers

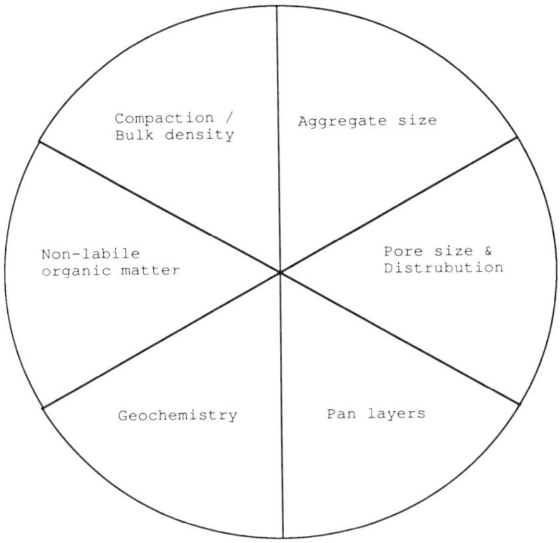

Figure 3. Temporally static characteristics of the rhizosphere.

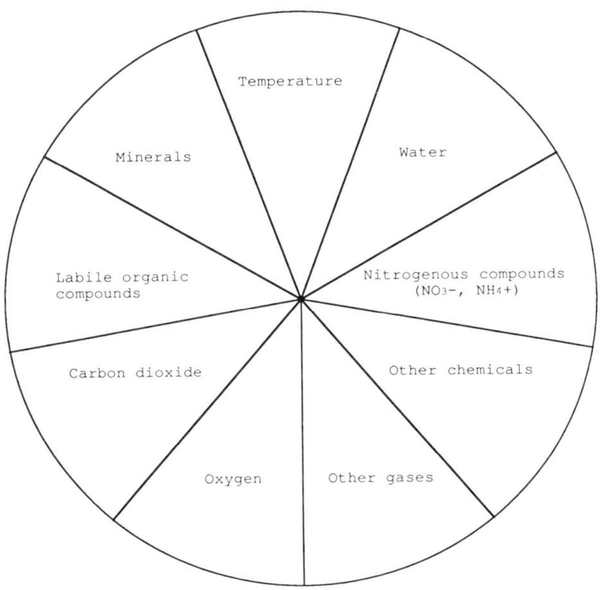

Figure 4. Temporally variable characteristics of the rhizosphere.

to several square meters or less. Characteristics which are temporally variable over a growing season (Fig. 4) can change over periods as short as minutes (temperature, rainfall) to as long as months (minerals). All rhizosphere environmental characteristics interact with each other to predicate their individual and joint temporal and spatial patterns. The challenge is to understand these interactions and their impact on the plant root system. More specifically, the challenge is to understand the developmental and physiological adjustments which plants can and do make in response to these edaphic characteristics and their temporal changes. This understanding must also include the effects of these plant adjustments on the edaphic characteristics.

Many excellent reviews of root soil interaction have been published over the last 30 years, and some of the older examples still remain the standards (Grable, 1966; Cooper, 1973). Some of the more recent reviews have done excellent jobs in synthesizing the knowledge obtained into usable generalizations (Russell, 1977; Harley and Russell, 1979; Miller, 1986). In this paper, therefore, I will look at some less well-documented plant responses to several of the temporal edaphic characteristics (carbon dioxide, temperature) and explore some of the implications of these interactions. I hope that the reader will better comprehend the inherent complexity of the root-soil system and will finish this paper with the opinion that not only is there much more to do, but that positive results toward relieving "constraints" are only a matter of time and effort.

B. Temperature

In 1973, A.J. Cooper made an in-depth review of root zone temperature and its effects on plant growth. The content of Cooper's review covers material from root zone temperature control of dry weight gain by plant parts and control of yield and morphological characteristics to techniques for the practical control of root zone temperature. Most of his conclusions are germane to much of the current research on low input, sustainable agriculture and alternative tillage practices. Root zone temperature affects crop growth in the field at nearly every location in the United States. We normally only consider an aspect of the soil as a factor in agriculture when the effect is a relatively deleterious one, i.e., reduction in yields, poor germination, heat-induced stress, etc. Most locations are concerned with low soil temperatures during germination, thus predicating the timing of planting, just as frost usually predicates the termination of the growing season and, therefore, growing season length. The impact of soil temperatures during the growing season is not normally of concern to the majority of temperate locations.

Growing degree day (GDD) calculations have been used extensively to predict the time of anthesis (Cross and Zuber, 1972) and many other aspects of plant phenology (Stevens et al., 1986). A recent study in New York has shown that the assumptions based on air temperature calculations of growing degree days may not be as applicable to seed germination in the field as soil GDD. Monk (1990) studied the emergence of corn cultivars under no-till and conventional-till. In estimating days to 50%

emergence, he found that the use of a base temperature of 10°C for 5 cm deep soil temperatures or 10°C for air temperatures gave GDD_{10} estimates which correlated to measured time of emergence more poorly than using a base temperature of 17°C (GDD_{17}) for 5 cm deep soil temperatures (R^2 = 0.86 and 0.91 vs. 0.97). He based his choice of the higher base temperature on the observation that corn roots do not show a linear increase in root growth with increasing temperatures until soil temperatures exceed 15–16°C (Brouwer, 1962; Grobbelaar, 1963; Cooper, 1973). This is consistent with the observations of Walker (1969) that a single degree change in root temperatures at these low temperatures results in a significant change in shoot response. In addition, data of Cal and Obendorf (1972) suggest a threshold temperature of 17–18°C for *normal* germination of corn.

Although it has been commonly assumed that soil temperatures at depths below two meters are constant throughout the growing season, field data do not support this assumption. Smith et al. (1964) display data which show that, at Ames, Iowa, the temperature at about two meters down in the soil profile changes from 5°C to 17°C over the course of a year, with a change from 9°C to 17°C over the period from May 1 to August 15. A similar pattern has been observed in upstate New York (Zobel, unpubl. data). Where the soil physical characteristics are suitable, for instance in Iowa, the roots of corn and soybean commonly grow to two meters or more in depth (Kaspar et al., 1984). The observations of Kaspar et al. (1984) suggest that corn (*Zea mays*) and soybean (*Glycine max*) roots follow the 16–17°C front down into the soil profile. These observations are consistent with similar conclusions by Stone and Taylor, (1983) based on a computer model and evidence from the literature.

In their review of tillage and soil environment effects on root growth, Allmaras et al. (1973) demonstrate that, early in the season, straw mulch suppresses the plow layer soil temperature by 1–2°C (on a 20–22°C base). This effect translates to the deeper portions of the soil as the season progresses. They cite evidence that, in the northcentral United States and southcentral Canada, this effect appears to reduce yields while, in the southeastern United States, it stimulates yields. In the latter case, this occurs by keeping soil profile temperatures in a more moderate range than would occur in these naturally warm climates. These treatments also impact moisture infiltration and retention to a significant extent. The moisture levels interact with the temperature effects to moderate both exceptionally low and high soil temperatures.

To what extent can these low temperatures at depths below the plow layer significantly affect crop yield? Consider the following research: Rykbost et al. (1975) obtained increases in corn dry matter yield of up to 45% and an increase of up to 105% for grain yields by heating the soil at 92 cm depth to 35°C. This heating resulted in an increase of only 1°C at 5 cm and 4°C at 25 cm where soil temperatures normally are in the range of 20–30°C. Temperature changes of 4–5°C when the root zone temperature is between 20 and 30°C have relatively little effect on the plant biomass (see Cooper, 1973). Because of their significantly warmer environment, the roots below the plow layer had an increased role in conditioning yield production. Similar results were obtained with a wide variety of crops (Rykbost et al., 1975). Other research suggests that there may be more practical methods of achieving an increase in deep soil temperatures (Allmaras et al., 1973; Cooper, 1973).

There is slight, but increasing, evidence that different types of roots have differential responses to the environment, including temperature (see Zobel, 1986, 1990a,b). Crop species, such as wheat and corn, place heavy reliance on basal and tap roots early in the season and on adventitious and lateral roots later in the season. For these species, breeding for cold tolerance in the seedling stage may have little relevance to the growth and development of the plant when the nodal (adventitious) roots are following the 17–18°C isotherm down into the soil. The observations of Allmaras et al. (1973) and Rykbost et al. (1975) suggest that breeding adventitious (nodal) roots of these species for cold tolerance may be highly advantageous.

C. Gases

The effect of reduced oxygen content on root and plant growth has been thoroughly studied (see Russell, 1977) and is essentially restricted to soils which are saturated or nearly saturated with water, resulting in a reduction of oxygen concentration ($[O_2]$) below 2%. In a plant, this low level of oxygen triggers a complex of responses which includes the production of ethylene, a gaseous plant growth regulator (Jackson and Campbell, 1975; Kawase, 1976; Yang, 1980). Many of the short-term effects reported for reduced oxygen can be directly traced to the effects of this growth regulator.

Ethylene gas is normally produced at extremely low levels by the plant during growth and development (10 ppb internal concentration; see Zobel and Roberts, 1978). It is involved in lateral root but not basal root initiation (Zobel, 1972, 1975) and is known to stimulate adventitious root initiation (see Abeles, 1973). At higher concentrations, ethylene inhibits root growth and root geotropic responses (Ycas and Zobel, 1983). In the field, this gas normally does not reach concentrations which would be of concern in root growth except under anaerobic conditions or when plants and/or roots meet severe physical stress (Abeles, 1973).

Carbon dioxide (CO_2) is present in the soil but its importance in root growth and subsequent shoot growth in the field has been discounted (Grable, 1966; Russell, 1977; Ycas and Zobel, 1983). Most of the arguments revolve around the high solubility of CO_2 in water. This solubility suggests that, under anaerobic conditions, this gas will not accumulate to toxic levels as a result of normal or even abnormal root function. In attempting to justify the use of aeroponics rather than hydroponics in root research, Zobel (1989) suggests several important roles or effects of carbon dioxide under field conditions. In greenhouse and growth chamber studies, the small size of rooting containers, absence of extensive microflora/fauna, and/or very porous nature of the media in which the roots are contained allow little accumulation of CO_2 in the rhizosphere of potted plants. On the other hand, under reasonably aerobic conditions in the field, concentrations of CO_2 can reach levels in excess of 7% (Evans, 1962; Buyanovsky and Wagner, 1983). Presumably, the difference in CO_2 levels is due primarily to rhizosphere size and quality (microorganisms). If CO_2 has a significant effect on plant roots, these differences would present a significant contrast between field and controlled-environment experimental results.

Minchin and Pate (1973) have estimated that 47% of concurrent plant photosynthate is fed to the rhizosphere through the roots. In this regard, it is interesting to note that 15% of the carbon dioxide fixed by corn plants in the field can come from the soil rather than the air (Allen, 1975). Since plants are the source of fixed carbon, these observations suggest an element of recycling of carbon through the rhizosphere. These estimates of the amount of carbon moved to the rhizosphere and the amount recycled also provide a conceptual basis for the rather high observed CO_2 contents within the soil where diffusion rates are considerably reduced from that of the open air. Soil physical and moisture characteristics and amounts of organic matter in the soil all modulate the concentration of carbon dioxide in the rhizosphere.

Research into the effect of carbon dioxide enrichment of the root zone has demonstrated many rather phenomenal results, not all of which are applicable to field conditions. Carbon dioxide concentrations up to 45% have been applied to the root system of crop plants, with concomitant increases in photosynthesis and root and shoot biomass (Erikson, 1946; Arteca and Dong, 1981; Arteca et al., 1979). Longer term studies demonstrate that these concentrations are inhibitory if continued beyond several hours or days (cf. Yurgalevitch and Janes, 1988). On the other hand, long-term treatment of the rhizosphere with CO_2 concentrations in the 0.1 to 0.5% range may show as much as 20 to 25% stimulation in growth compared to ambient CO_2 levels (Geisler, 1969; Yurgalevitch and Janes, 1988).

Excised root culture studies have demonstrated that carbon dioxide concentrations above ambient atmospheric levels stimulate root growth (Talbot and Street, 1968). Mineral nutrition is also responsive to carbon dioxide concentration with 20 to 25% increases in uptake as CO_2 concentrations reach the 0.5 to 1.0% level (Yurgalevitch and Janes, 1988). Unfortunately, detailed research on the effect of carbon dioxide enrichment of the rhizosphere is spotty and generally deals only with concentrations at the high end—greater than 2%. There is a need for research to document the effects of CO_2 enrichment of the rhizosphere in the 0.05 to 1.0% range. The above discussion suggests that global climate changes (as projected by some weather models) and the use of tillage techniques which increase soil organic matter or suppress diffusion out of the soil may very well result in significant increases in soil CO_2. This possibility and its consequences must be assessed.

III. Root System Morphology

While much of the foregoing is probably familiar to most readers, it may not be necessarily interpreted in this fashion. On the other hand, the complexity of the root system is frequently overlooked. A series of recent papers in symposia and review books have addressed our knowledge of plant roots (Torrey, 1976; Miller, 1986; Zobel, 1986, 1989, 1990, 1991a, 1991b; Gregory et al., 1987; Waisel et al., 1991; see also Klepper, this volume). The situation is not as clear as once believed. Most of the physiological and soils research undertaken in recent decades have relied on accounts of root morphology and anatomy which were summarized best by Esau (1965). Statements about roots in this classic text were based on the accumulation of knowledge over the previous 100 to 150 years. With improving technology and research into the physiological function, anatomical/morphological development, and genetic control of roots, these conclusions are being demonstrated to be gross over-generalizations about an extremely complex biological system. The root system must be thoroughly understood if consistent advances are to continue to be made in agriculture.

The most significant recent discoveries suggest a lack of functional similarity between roots of different types on the same plant. McCully and others in her laboratory (St. Aubin et al., 1986; McCully, 1987; Kevekordes et al., 1988; McCully and Canny, 1989) have demonstrated that, under field conditions, the adventitious nodal (prop) roots of corn do not provide a significant proportion of the concurrent water and mineral needs of the plant. These are, of course, the roots which penetrate to the deepest layers of the soil profile and provide the structural framework for the overall root system. These roots simply do not have, within their apical half, the mature xylem vessel members needed for the transport of significant amounts of water and nutrients. At the same time, the distal half of these roots has mature xylem vessels and an intact endodermal casparian strip. In corn, this phenomenon may hold only for these adventitious nodal roots; lateral roots appear to have mature xylem vessels developing only a short distance behind the zone of root elongation. A similar phenomena occurs with barley (*Hordeum vulgare*) (Sanderson et al., 1988), soybean, and presumably other dicots in their secondary vascular tissues, though it may not be nearly as severe (Kevekordes et al., 1988).

Zobel (1975, 1980, 1986, 1991a) has referred to four distinctly different types of roots and suggested the possibility of additional ones. McCully (1987) has described the anatomy of a class of very small corn lateral roots as having a rudimentary vascular system with a thin cortex. Fusseder (1987) and, later, Cahn et al. (1989) described the smallest roots of corn as being determinate in growth habit. These are probably the same roots as discussed for corn by McCully (1987) and for barley by Hackett and Rose (1972). With their reduced anatomy and short growing span (two days and a life span of two to four weeks), these roots are not likely to have the same functional characteristics as larger lateral and nodal roots. With the increasing understanding of the functional and structural differentiation of roots (see Zobel, 1986; Waisel and Brekle, 1987; Waisel et al., 1991), a consistent nomenclature system is needed to prevent confusion of root research results (cf. Zobel, 1991a, for a discussion of this need and a proposed nomenclature system).

IV. Root-Soil Interactions

Briefly, a given edaphic environment is the result of a multitude of interactions between the many aspects of the soil. Carbon dioxide levels of the soil are the direct result of the respiration of roots and soil micro- and macro-organisms. The moisture and temperature levels in the soil further modulate carbon dioxide levels by affecting the activity of these organisms. They also modulate these levels through the physico-chemical controls on diffusion and solubility. Carbon dioxide concentrations are further modulated by pore size and distribution as well as the geometry of the soil surface. These predominantly temporal factors of the edaphic atmosphere have dramatic interactions with the spatial factors characterized by soil physical and chemical characteristics as well as the roots and micro- and macro-organisms. Spatial variability across distances of less than a few meters to several kilometers and from a few centimeters to several meters in depth from the surface can dramatically change these interactions for a given site. These interrelationships between different soil parameters are extensive and effectively control all aspects of root and shoot growth.

A. Patterns of Interaction

Patterns of root response to any one or a suite of these edaphic factors are not fixed. Different genotypes and cultivars within the same species have their own characteristic response patterns when presented with a given or temporally changing edaphic environment (Zobel, 1975). These

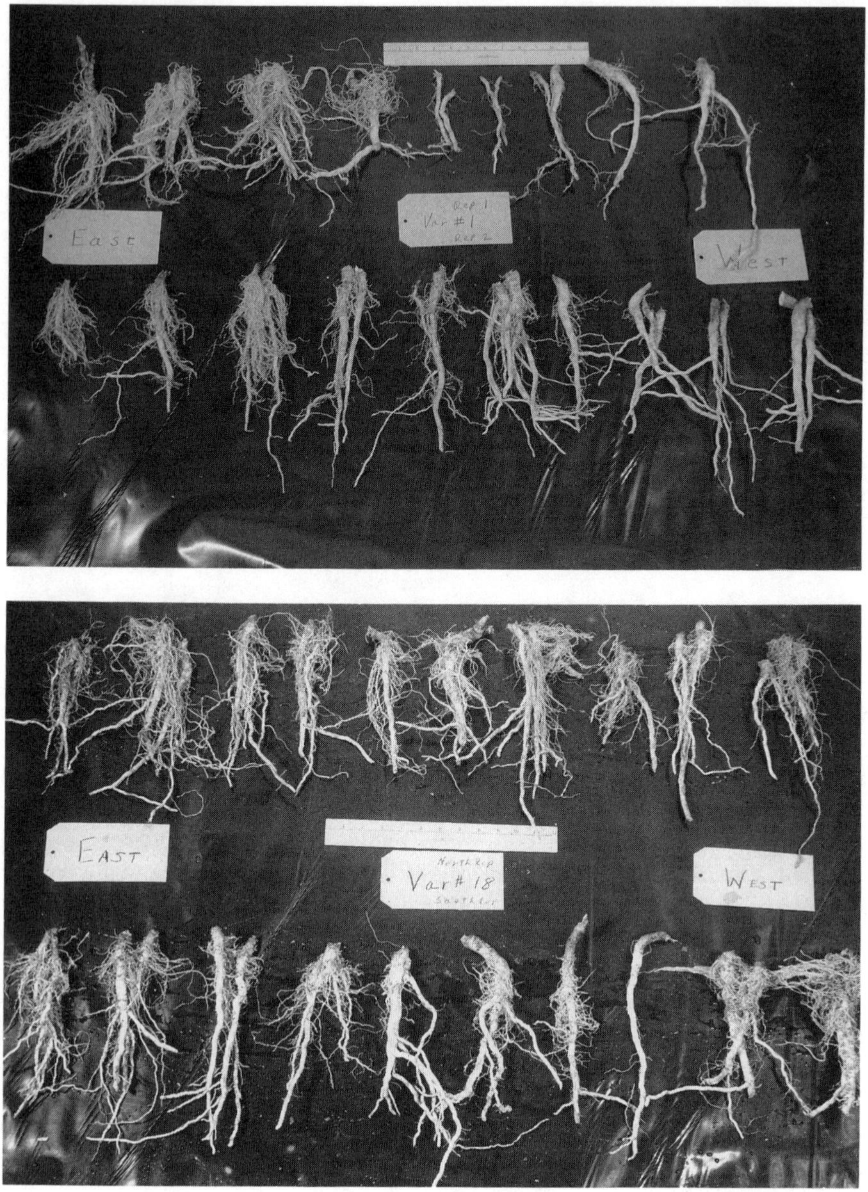

A

B

response patterns can lead to misleading conclusions when only a few genotypes or sites are considered. Consider the contrast between the root system morphologies and response patterns of three different tomato (*Lycopersicon esculentum* Mill.) genotypes within several meters of each other in the field (Fig. 5). Two genotypes (cv. VF13L and VF65C)

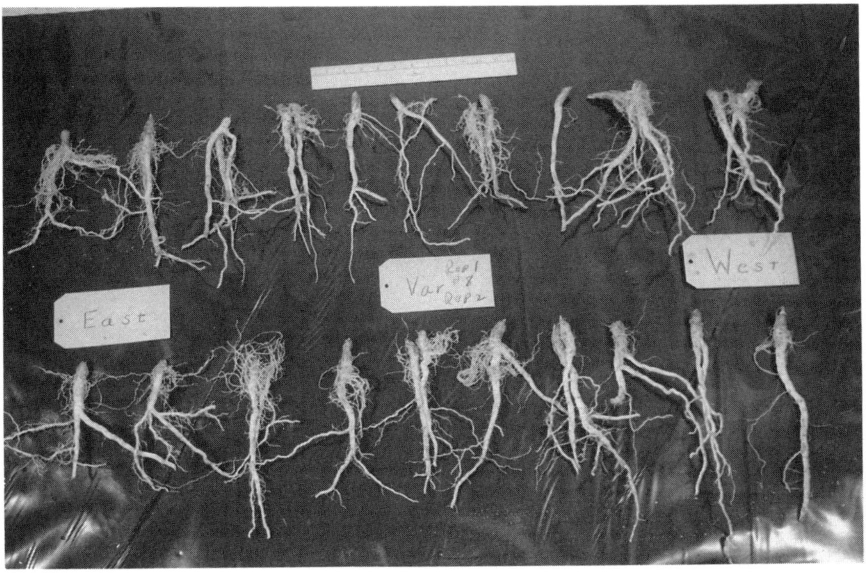

C

Figure 5. Roots from two reps each of three tomato cultivars (A = VF145-78-79; B = VF65C; C = VF13L-34) grown in half-mile rows at the Bob Button farm in Winters, CA, 1971. The soil is a Yolo sandy loam (Typic Xerorthents; fine-silty, mixed, non-acid thermic) on the west end of the field grading into a Yolo sandy clay (Typic Chromoxererts; fine, montmorellenite) towards the east end of the field. These replications were part of a randomized experiment. The roots were excavated with a shovel of soil and represent the root system present in the plow layer.

demonstrate no change as the soil grades from a sandy loam to a sandy clay; however, one (VF65C) has large numbers of fine roots along the whole row and the other (cv. VF-13L) has few along the whole row. In contrast, the third cultivar (VF145-78-79) demonstrates a change in morphology with few fine roots (similar to VF13L) in sandy loam soil and many fine roots (similar to VF65C) in sandy clay soil. Similar results were found with several other cultivars in the same experiment (Zobel, 1975; Smith and Zobel, 1991) and in a follow-up experiment by Gulmon and Turner (1978). Corn also demonstrates similar responses when the roots encounter differential rooting depths (Fig. 6).

The differential morphological response of cultivars to given soil environments has been termed phenotypic plasticity. O'Toole and Bland (1987) briefly review some of the evidence for and significance of phenotypic plasticity. Smith and Zobel (1991) refer to morphoplastic and morphostatic phenotypes. In these terms, the roots of the cultivars VF13L and VF65C in Figure 5 are morphostatic and the others are morphoplastic under the conditions presented. The cultivar VF13L is

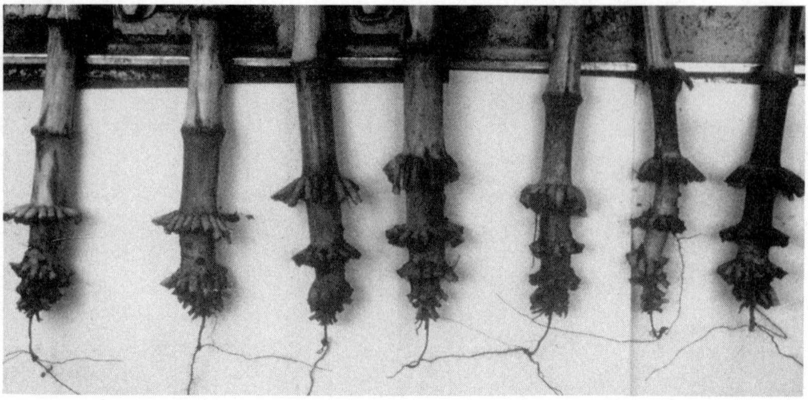

Figure 6. Roots from a single rep of corn (cv. ASGROW RX788) sampled sequentially every 10 meters along an 80-meter row. The plants were grown at the Rodale Research Center as part of the Farming Systems Trial—conventional treatment in rep 4 in 1987. The soil was a Comly silt loam (Typic Fragiudalf; fine-loamy, mixed, mesic [Matelski, 1976]) with an average rooting depth of 30–45 cm on the east end changing to an average rooting depth of 45–60 cm on the west end of the row (right hand side of the figure).

probably physioplastic (i.e., it can modify its physiology rather than its morphology to adapt to the changing rhizosphere environment) while, by contrast, the others are physiostatic, VF145-78-79 suffered a reduced yield where it had increased numbers of fine roots and VF65C had much reduced yields throughout the field compared to expectations from other yield trial experiments (Zobel, 1971). The differential tendency to modify the morphology of the root system in response to environmental differences or changes is a severely complicating factor in the analysis of root/soil interaction.

Our inability to deal effectively with this maze of interactions encompassing both spatial and morphological variability has resulted in the use of uniform laboratory conditions and relatively uniform research fields as the sites for root research and studies on the constraints to root growth. Modern science is predominantly deductive in design and analysis. This deductive approach requires experiments where the number of changing variables is reduced to an absolute minimum. This allows the observation that when X (condition, level, etc.) is changed then Y response is obtained. By linking these results to others, more complex experiments may be obtained with some hope for understanding the results. This reductionist approach can be successful with situations where up to three or four parameters are interacting, but more complex situations make the experimental design unwieldy and/or too expensive in terms of manpower and materials. One reason for this reductionist approach has been the

Table 1. Table of percentages which certain partitions of SS have of the SS from which they were partitioned. (Extracted from Zobel, 1990)

Character	Percentage of SS (Ratio* 100)	
	TRT/TOT	GxE/TRT
Total roots	74.6	66.1
Lateral roots	72.3	65.8
Tap diam.	74.3	64.3
Hypocotyl diam	75	55.5
Basal roots	66.4	53.6
Basal diam.	61.2	52.3
T-6 diam	66.6	47.3
Root biomass	75.6	44.3
Shoot biomass	76.6	32.6
Vegetat. stage	87.2	28.6
Yield	91	24.4
Height	95	16.1
Reproduct. stage	93.6	14.6

lack of suitable procedures to characterize the relative importance of differing factors in a complex interactive experiment or in a natural environment.

B. Statistical Analysis of Interaction

Because plant breeding efforts have focused on the shoot rather than the root, root characteristics have a much higher level of interaction variance than other plant characteristics (Table 1). As much as two-thirds of the treatment variance is due to interaction with the environment (Table 1; see also Zobel, 1991a). With this level of interaction and no adequate method to analyze it, little progress is possible. Suitable procedures have been in existence for quite some time, but only the advent of the computer has made them available to the scientist and statistician interested in the more complex environments found in the soil. Zobel (1990) has described a procedure (Additive Main Effects and Multiplicative Interaction—AMMI; see Figs. 7, 8, 9 for programs to effect an AMMI analysis in two commercial statistical packages) for studying genotype by environment interaction and, by extension, environment by environment interaction.

Zobel (1991a) points out that, of the many difficulties impacting research on the constraints to root growth, the lack of clear delineation of the type of root being studied causes the most severe confusion. Figure 10, for example, presents an interaction biplot (means plotted on the x-axis and scores for interaction axis-1 plotted on the y-axis) of the

AMMI analysis with SAS.

AMMI is suitable for two-way data sets or data sets which can be constrained to be, in effect, two-way data sets. In the following, Geno refers to genotype or more generally to the variables for one side of the two-way matrix and Env (environment) refers to the other side of the two-way matrix. Any other names can be used, but for convenience and clarity these two are ones used here. It is also assumed that geno is the shorter set of variables.

In the following listing of a SAS program you will note that GLM (ANOVA will work as well) is run twice. The first time is to get the ERROR df, SS, and MS correct for an ANOVA with interaction. The second run is to get the residuals and run the complete AMMI analysis. The shortest set of variables **MUST** be used in lieu of GENO and listed in the VAR statement of PROC PRINCOMP.

```
DATA;
    INFILE SUBSET;
    INPUT ENV $ GENO $ REP $ Y;
PROC SORT;
    BY GENO ENV;
PROC GLM;
    CLASS GENO ENV;
    MODEL Y=GENO ENV GENO*ENV;
PROC GLM;
    CLASS GENO ENV;
    MODEL Y=GENO ENV;
    OUTPUT OUT=SUBRES R=RESID;
    MEANS GENO ENV;
PROC MEANS DATA=SUBRES NOPRINT;
    VAR RESID;
    BY GENO ENV;
    OUTPUT OUT=RESMEAN MEAN=RMEAN;
PROC SORT;
    BY ENV GENO;
PROC TRANSPOSE DATA=RESMEAN OUT=PCA;
    VAR RMEAN;
    ID GENO;
    BY ENV;
PROC PRINCOMP COV DATA=PCA OUT=PCAOUT;
    VAR G40 G60 G180 G853 G890;
PROC PRINT;
    BY ENV;
    VAR PRIN1 PRIN2 PRIN3;
```
----

*Mention of a trademark or proprietary product does not constitute a guarantee or warranty of the product by the USDA and does not imply its approval to the exclusion of other products that may also be suitable.

Figure 7. Description and program by which an AMMI analysis can be accomplished using the SAS* statistical package.

data which demonstrate that site specific characteristics control total root numbers in soybeans grown in New York State. Most studies of root growth in the field use this same type of data—total root numbers irrespective of root type or even size. There is, therefore, no delineation as to any differences in root function, longevity, etc. The futility of such analyses is clearly demonstrated in Figure 11 where a subset of the total root data is analyzed separately. This biplot demonstrates that, for basal roots in soybeans, it is temporal characteristics, as characterized by

To convert SAS output to standard AMMI values:

Assume: G= number of genotypes; E= number of environments; R= number of Replicates; df = degrees of freedom for a given axis; n = axis number; SS1 = sums of squares for axis 1; Ev1 = Eigen-value for axis 1; Eg1 & Ee1 = genotype and environment, respectively, eigenvectors for axis 1; Sg1 & Se1 are the respective scores. Subscripts indicate the source of the data used in the calculations.

$df_n = E + G - 1 - 2n$

the following are for axis 1. The procedure for other axes is identical, using the appropriate values.

$SS1_{AMMI} = Ev1_{SAS} * (E-1) * R = SS$ for IPCA 1
$Ev1_{AMMI} = Ev1_{SAS} * (E-1)$
$Eg1_{AMMI} = Eg1_{SAS}$
$Sg1_{AMMI} = Eg1_{SAS} * (Ev1_{AMMI})**0.25$
$Ee1_{AMMI} = Se1_{SAS} / (Ev1_{AMMI})**0.5$
$Se1_{AMMI} = Ee1_{AMMI} * (Ev1_{AMMI})**0.25$

The sum of SS_{AMMI} across all possible axes is the GENO*ENV SUM OF SQUARES and the sum of the df for all possible axes is the df for the GENO*ENV source of variance.

AMMI ANOVA table for BASAL root data

Source	df	SS	MS	Probability
Total	599	7835.7	13.08	
TRT	199	5203.7	26.14	0.000 ***
GENO	9	408.2	45.35	0.000 ***
ENVR	19	2003.1	105.42	0.000 ***
G X E	171	2792.4	16.32	0.000 ***
IPCA 1	27	708.1	26.22	0.000 ***
Residual	144	2084.2	14.47	0.000 ***
Error	400	2632.0	6.58	

AMMI scores are scaled to be valued as the square root of the dimensions of the mean values for each treatment, i.e. if the means are in Kg/Ha then the scores are (Kg/Ha)**0.5. This simplifies the interpretation of biplots and allows estimation of predicted treatment means according to the AMMI model:

$$Y_{ge} = \mu + \alpha_g + \beta_e + \sum_{n=1}^{N} \lambda_n \gamma_{gn} \delta_{en} + \theta_{ge}$$

----

*Mention of a trademark or proprietary product does not constitute a guarantee or warranty of the product by the USDA and does not imply its approval to the exclusion of other products that may also be suitable.

Figure 8. Method for converting SAS* output into standard AMMI values.

annual events, rather than spatial (site-specific) characteristics which control the interaction between basal root numbers and the environment. Which specific temporal and spatial factors are involved in these interactions is currently being explored (Bartlett, 1978a, Zobel, 1990).

Spatially induced variation within an experimental site causes an increase in the error variance in an Analysis of Variance (ANOVA). Proper experimental and plot design can frequently reduce this variance. Several statistical procedures, such as randomized complete block,

```
OPEN 'GEN.DAT';CHANNEL=2;FILETYPE=INPUT
    UNITS [NVALUES=75]
    FACTOR [LEVELS=3;VALUES=25(1...3)] REPS
      &  [LEVELS=5; LABELS=!T(E841,E849,E856,E861,E875)] ENVIRON
      &  [LEVELS=5; LABELS=!T(G40 ,G60 ,G180,G853,G890)] GENOTYPE
    VARIATE BASAL
READ [CHANNEL=2;PRINT=*] ENVIRON, GENOTYPE, REPS, BASAL;\
       FREPRESENTATION=LABELS,LABELS,*,*
    TREATMENTSTRUCTURE GENOTYPE*ENVIRON
ANOVA [PRINT=*] BASAL;SAVE=AOV1
    TREATMENTSTRUCTURE GENOTYPE + ENVIRON
ANOVA [PRINT=*] BASAL; RESIDUALS=RESID; SAVE=AOV
    AKEEP GENOTYPE + ENVIRON; MEANS=GMEANS,EMEANS
  TABULATE [CLASSIFICATION=GENOTYPE,ENVIRON] RESID;MEANS=MRES
    FACTOR [LEVELS=5;LABELS=!T(AXIS1,AXIS2,AXIS3,AXIS4,AXIS5)] AXIS
    MATRIX [ROWS=5;COLUMNS=5] RESM
    MATRIX [ROWS=GENOTYPE;COLUMNS=AXIS] GEIG
    MATRIX [ROWS=ENVIRON;COLUMNS=AXIS] EEIG
    MATRIX [ROWS=GENOTYPE;COLUMNS=AXIS] GSCORE
    MATRIX [ROWS=ENVIRON;COLUMNS=AXIS] ESCORE
    DIAGONALMATRIX [ROWS=5] SING
  EQUATE OLDSTRUCTURE=MRES; NEWSTRUCTURE=RESM
SVD RESM; LEFT=GEIG;SINGULAR=SING;RIGHT=EEIG
  CALCULATE SRTS=SQRT(SING)
  CALCULATE GSCORE = GEIG*+SRTS
  CALCULATE ESCORE = EEIG*+SRTS
  CALCULATE INTERSS = SING*SING*3
    MATRIX [ROWS = GENOTYPE; COLUMNS = 1] GIPCA1
    MATRIX [ROWS = ENVIRON; COLUMNS = 1] EIPCA1
  CALCULATE GIPCA1=GSCORE$[!(1...5);!(1)]
  CALCULATE EIPCA1=ESCORE$[!(1...5);!(1)]
    ADISPLAY [FPROBABILITY=Y;PRINT=AOVTABLE] SAVE=AOV1
    ADISPLAY [FPROBABILITY=Y;PRINT=AOVTABLE] SAVE=AOV
    PRINT GMEANS,EMEANS,INTERSS,GSCORE,ESCORE
    GRAPH Y=GIPCA1,EIPCA1;X=GMEANS,EMEANS;SYMBOLS=!T(G1,G2,G3,G4,G5),\
      !T(E1,E2,E3,E4,E5)
```

This program produces an ANOVA table for interaction and for non-interaction models; a Sums of Squares table with values for each interaction axis; and tables of interaction scores for all genotypes and environments. In addition it produces a biplot of the first axis.

----

*Mention of a trademark or proprietary product does not constitute a guarantee or warranty of the product by the USDA and does not imply its approval to the exclusion of other products that may also be suitable.

Figure 9. Program for a complete AMMI analysis using the Genstat statistical package.

attempt to move some of this spatial variance from the error to the treatment variance. Unfortunately, none of these designs assess all of the spatial variance, though some do assess the interaction between the captured spatial variance and some of the treatment variance (Bartlett, 1978a, Zobel et al., 1988).

Zobel has recently suggested that the procedure he uses for interaction analysis is a restrictive subset of the more general Principal Components Analysis (PCA) procedure (Zobel, personal communication). He further suggests that coupling PCA analysis of the spatial characteristics of a field experiment with the AMMI analysis of the treatment effects should allow for more precise interpretations of the bases for interactions. Eisenberg et al. (in prep.) apply this theory in a classical nitrogen by genotype/by

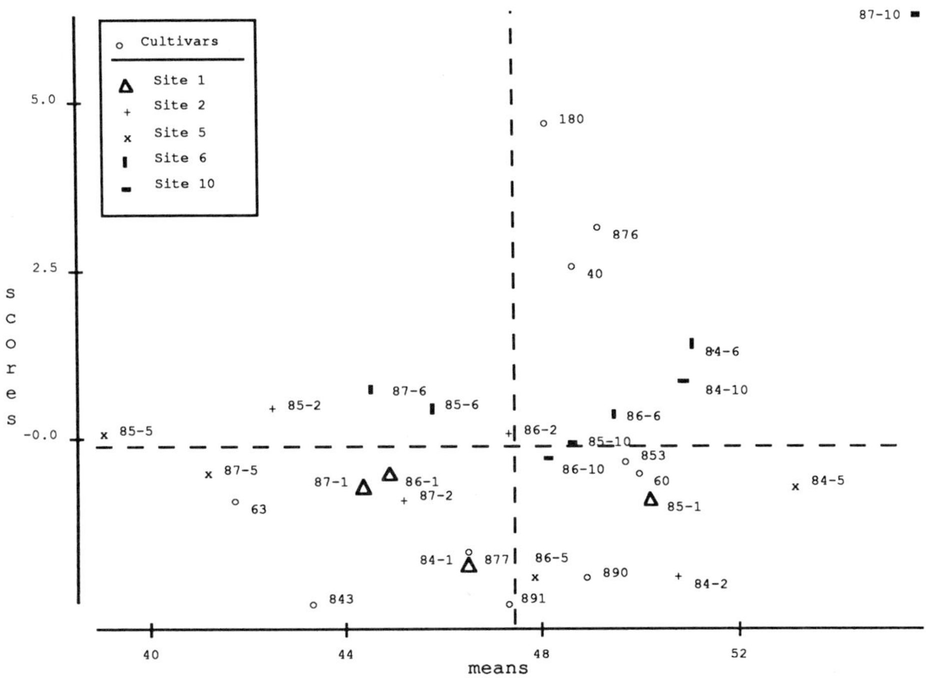

Figure 10. Biplot of the first AMMI interaction axis scores (y-axis) plotted against mean numbers of total roots (x-axis) for TOTAL NUCLEAR ROOT NUMBERS of ten soybean cultivars grown at five sites over four years (20 site-years). The analysis used the procedures described in Figures 8 and 9. (84 = 1984; 85 = 1985; 86 = 1986; 87 = 1987; −1 = site 1 [Aurora, NY]; −2 = site 2 [Mt. Morris, NY]; −5 = site 5 [Canton, NY]; −6 = site 6 [Chazy, NY]; −10 = site 10 [Ithaca/Varna, NY]; see Zobel et al. 1988; Zobel, 1990, for details on materials, methods, and cultivar names.)

site/by year study with wheat. By plotting the residuals from an ANOVA of the data from one site on a field map, they demonstrated that there was a strong spatial pattern to these residuals. This pattern was removed by applying the AMMI analysis to the residuals ordered in a row by column arrangement based on the map rather than by treatments. This process partitions the field spatial variability into a separate term, decreasing the size of the error Mean Square (MS) and thus increasing the precision of the analysis. Similar approaches using nearest neighbor analyses and other spatial analysis procedures have been proposed (Papadakis, 1937; Bartlett, 1978a,b). Testing of the efficacy of these procedures and their relative merits and demerits should provide root/soil scientists with a suite of powerful tools to explore interactions.

While providing a powerful aid to classical experimentation and its deductive reasoning approach, these new statistical procedures are even more efficient for inductive reasoning approaches where holistic studies

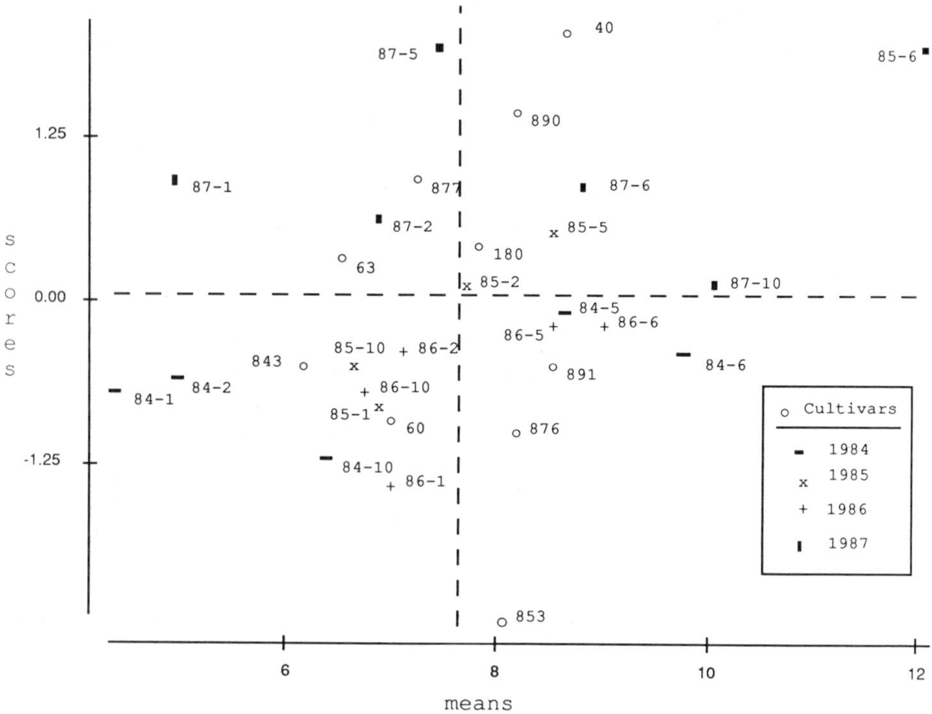

Figure 11. Biplot of the first AMMI interaction axis scores (y-axis) plotted against mean numbers of total roots (x-axis) for BASAL ROOT NUMBERS of ten soybean cultivars grown at five sites over four years (20 site-years). The analysis used the procedures described in Figures 8 and 9. (84 = 1984; 85 = 1985; 86 = 1986; 87 = 1987; −1 = site 1 [Aurora, NY]; −2 = site 2 [Mt. Morris, NY]; −5 = site 5 [Canton, NY]; −6 = site 6 [Chazy, NY]; −10 = site 10 [Ithaca/Varna, NY]; see Zobel et al. 1988; Zobel, 1990 for details on materials, methods, and cultivar names.)

with complex environments are analyzed. Because of the inherently complex nature of the rhizosphere, classical experimental approaches are only marginally able to dissect the complex patterns of the interactions between the soil and plant roots.

V. Conclusions

At first glance, it does not appear that the complaint of the first sentence of this manuscript has been adequately addressed. This is done here, now that the necessary background material has been introduced. A given cultivar of a specific species, grown at a single location in a single year, may indeed have soil environment constraints to root growth and development. As mentioned in the text, some environmental parameters not routinely associated with being a constraint (CO_2, deep soil temperatures, ethylene gas) can further restrict or modify growth. The total set of environmental characteristics of the soil interact with each other to dictate the exact edaphic environment upon which plant roots depend at a given point in time. If the level of just one parameter changes, a number of characteristics change in response, resulting in a new environment based on a complex series of interactions. The edaphic environment is, therefore, in a dynamic rather than static state, with equilibria which may be very short-lived. The understanding of the edaphic environment, which is critical to an understanding of root/soil relationships, can not take place without an adequate grasp of these complex interactions or, at least, a holistic view of their dynamic nature.

On the other side, plants have adapted to complex edaphic environments over millions of years while breeders have been cryptically modifying these adaptations for only a few thousand of years at best. Since most of this selection has taken place on elite soils, few of the adaptations have been severely modified. This results in a very wide pattern of response to a given edaphic environment by a set of elite cultivars. This diverse response pattern becomes even more extreme when non-elite soils and a wide set of cultivars are used. Ultimately, it can be suggested that, within reason, a genotype adapted to nearly any given edaphic environment can be found. This hypothesis then suggests that soil environments do not, per se, constrain root growth. The constraint is more in terms of knowledge of the interactions and the availability of germplasm adapted to a given environment. Documentation and categorization of the soil/soil and soil/plant interactions are necessary if advances in improving root soil interaction are to succeed.

The recent development of new statistical tools to handle interaction and the upsurge in research into root growth on an anatomical and morphological level presages rapid advances into this documentation and categorization. The end result will be holistic views of the root-soil interface which will allow rapid cultivar development and development of agronomic methods for most environmental conditions where agriculture is possible. The above points lead to six research challenges for the 1990's:

(1) To develop interdisciplinary teams (including modelers) to study the integration of edaphic and root characteristics (including micro- and macroorganisms) and produce a holistic view of the rhizosphere.
(2) To document how these plant responses in turn modify the edaphic characteristics with which they interact.
(3) To determine which edaphic characteristics (both quantitative and qualitative) plant root systems are actually adapted to and the mechanism(s) of that adaptation.

(4) To develop a thorough understanding of the genetic, developmental, and physiological controls which predicate morphostatic vs. morphodynamic and physiostatic vs. physiodynamic responses to edaphic conditions.
(5) To modify the soils and tailor them to the needs of the plants as an interim measure or, occasionally, as a permanent fix.
(6) To directly breed plants to manipulate root adaptation to problem soils.

References

Abeles, F.B. 1973. *Ethylene in Plant Biology*. Academic Press, London, 302 pp.

Allen, L.H. 1975. Line-source carbon dioxide release. III. Predictions by a two-dimensional numerical diffusion model. *Boundary-Layer Meteorology* 8:39–79.

Allmaras, R.R., A.L. Black, and R.W. Rickman. 1973. Tillage, soil environment, and root growth. In: *Proceedings of the National Conservation Tillage Conference*. Des Moines, Iowa, March 28–30, 1973, pp. 62–86.

Arteca, R.N., and C.N. Dong. 1981. Increased photosynthetic rates following gibberellic acid treatments to the roots of tomato plants. *Photosyn. Res.* 2: 243–249.

Arteca, R.N., B.W. Pooviah, and O.E. Smith. 1979. Changes in carbon fixation, tuberization, and growth induced by CO_2 applications to the root zone of potato plants. *Science* 205:1279–1280.

Bartlett, M.S. 1978a. Further analysis of spatial patterns: A re-examination of the Papadakis method of improving the accuracy of randomized block experiments. *Suppl. Advances in Applied Probability* 10:133–143.

Bartlett, M.S. 1978b. Nearest neighbor models in the analysis of field experiments. *J. Royal Statistical Society* B40:147–174.

Brouwer, R. 1962. Influence of temperature of the root medium on the growth of seedlings of various crop plants. *Jaarv. Inst. Biol. Scheik. Onderz. LandbGewass. Wageningen*, The Netherlands, p. 11–18.

Buyanovsky, G.A., and G.H. Wagner. 1983. Annual cycles of carbon dioxide level in soil air. *Soil Sci. Soc. Am. J.* 47:1139–1145.

Cahn M.D., R.W. Zobel, and D.R. Bouldin. 1989. Relationship between root elongation rate and diameter and duration of growth of lateral roots of maize. *Plant Soil* 119:271–279.

Cal, J.P., and R.L. Obendorf. 1972. Differential growth of corn (*Zea mays* L.). Hybrids seeded at cold root zone temperatures. *Crop Sci.* 12:572–575.

Cooper, A.J. 1973. *Root Temperature and Plant Growth*. Research Review No. 4. Commonwealth Bureau of Horticulture and Plantation Crops, Slough SL2 3BN, England, 73 pp.

Cross, H.Z., and M.S. Zuber. 1972. Prediction of flowering date in maize based on different methods of estimating thermal units. *Agron. J.* 64:351–355.

Erickson, L.C. 1946. Growth of tomato roots as influenced by oxygen in the nutrient solution. *Am. J. Bot.* 33:551–556.

Esau, K. 1965. *Plant Anatomy*. John Wiley and Sons, Inc., New York, pp. 481–538.

Evans, D.W. 1962. *Influences of oxygen and carbon dioxide in the rooting medium on the growth of seedling legumes*. Dissertation, Department of Agronomy, Cornell University, February, 1962.

Fusseder, A. 1987. The longevity and activity of the primary root of maize. *Plant Soil* 101:257–265.

Geisler, G. 1969. Die wirkung der kohlenasurekonzentration (CO_2 ± HCO_3) im bereich niedriger sauerstoffkonzentrationen (O_2) in der bodenluft auf das spross- und wurzel-wachstum von kulturpflanzen. *Z. Acker. Pflanz.* 130:1–15.

Grable, A.R. 1966. Soil aeration and plant growth. *Adv. Agron.* 18:57–106.

Gregory, P.J., J.V. Lake, and D.A. Rose. 1987. *Root Development and Function*. Cambridge University Press, London.

Grobbelaar, W.P. 1963. Responses of young maize plants to root temperatures. *Meded. LandbHogesch. Wageningen* 63:1–71.

Gulmon, S.L., and N.C. Turner. 1978. Differences in root and shoot development of tomato (*Lycopersicon esculentum* MILL). *Plant Soil* 49: 127–136.

Hackett, C., and D.A. Rose. 1972. A model of the extension and branching of a seminal root of barley, and its use in studying relations between root dimensions. Part II. Results and inferences from manipulation of the model. *Aust. J. Biol. Sci.* 25:681–690.

Harley, J.L., and R.S. Russell. 1979. *The Soil-Root Interface*. Academic Press, London, 448 pp.

Jackson, M.B., and D.J. Campbell. 1975. Movement of ethylene from roots to shoots, a factor in the responses of tomato plants to waterlogged soil conditions. *New Phytol.* 74:397–406.

Kaspar, T.C., H.M. Taylor, and R.M. Shibles. 1984. Taproot elongation rates of soybean cultivars in the glasshouse and their relation to field rooting depth. *Crop Sci.* 24:916–920.

Kawase, M. 1976. Ethylene accumulation in flooded plants. *Plant Physiol.* 36:236–241.

Kevekordes K.G., M.E. McCully, and M.J. Canny. 1988. Late maturation of large metaxylem vessels in soybean roots: Significance for water and nutrient supply to the shoot. *Ann. Bot.* 62:105–117.

Matelski, R.P. 1976. *Soil Series of Pennsylvania—Catena Diagrams*. Agronomy Series 43. Department of Agronomy, Pennsylvania State University, University Park.

McCully, M.E. 1987. Selected aspects of the structure and development of field-grown roots with special reference to maize. *In* P.J. Gregory, J.V. Lake, and D.A. Rose (Eds.). *Root Development and Function*. Cambridge University Press, London, pp. 53–70.

McCully, M.E., and M.J. Canny. 1989. Pathways and processes of water and nutrient movement in roots. *In* Loughman et al. (Eds.). *Structural and Functional Aspects of Transport in Roots*. Kluwer Academic Publishers, Dordrecht, The Netherlands, pp. 3–14.

Miller, D.E. 1986. Root systems in relation to stress tolerance. *Hort. Sci.* 21: 963–970.

Minchin, F.R., and J.S. Pate. 1973. The carbon balance of a legume and the functional economy of its root nodules. *J. Exp. Bot.* 24:259–271.

Monk, R. 1990. *Effects of environment and tillage systems on the establishment, development, and yield of six short season corn hybrids in northern New York*. Dissertation, Department of Agronomy, Cornell University, May, 1990, 244 pp.

O'Toole, J.C., and W.L. Bland. 1987. Genotypic variation in crop plant root systems. *Adv. Agron*. 41:91–145.

Papadakis, J.S. 1937. Methode statistique pour des experiences sur champ. *Bull. Institut Amel. Plantes Salonique* 23.

Russell, R.S. 1977. *Plant Root Systems: Their Function and Interaction with the Soil*. McGraw Hill, London, 298 pp.

Rykbost, K.A., L. Boersma, H.J. Mack, and W.E. Schmisseur. 1975. Yield response to soil warming: Agronomic crops. *Agron. J*. 67:733–738.

St. Aubin, G., M.J. Canny, and M.E. McCully. 1986. Living vessel elements in the late metaxylem of sheathed maize roots. *Ann. Bot*. 58:577–588.

Sanderson, J., F.C. Whitbread, and D.T. Clarkson. 1988. Persistent xylem cross-walls reduce the axial hydraulic conductivity in the apical 20 cm of barley seminal root axes: Implications for the driving force for water movement. *Plant Cell Env*. 11:247–256.

Smith, G.D., F. Newhall, L.H. Robinson, and D. Swanson. 1964. Soil-temperature regimes: Their characteristics and predictability. *USDA-SCS Publication*: SCS-TP-144, April, 1964. 14 pp.

Smith, M.E., and R.W. Zobel. 1991. Plant genetic interactions in alternative cropping systems: Considerations for breeding methods. In: Plant Breeding and Sustainable Agriculture: Considerations for Objectives and Methods. Crop Sci. Soc. Am. Special Publication 18, pp. 57–81.

Stebbins, G.L. 1966. *Processes of Organic Evolution*. Prentice-Hall, Englewood Cliffs, New Jersey, 191 pp.

Stevens, E.J., K.M. Eskridge, S.J. Stevens, A.D. Flowerday, and C.O. Gardner. 1986. Phenology of dent corn and popcorn. I. Analysis of repeated measurements from phenology experiments. *Agron. J*. 78:1081–1088.

Stone, J.A., and H.M. Taylor. 1983. Temperature and the development of the taproot and lateral roots of four indeterminate soybean cultivars. *Agron. J*. 75:613–618.

Talbot, B., and H.E. Street. 1968. Studies of the growth in culture of excised wheat roots. VII. Enhancement of growth by an unidentified volatile product of root metabolism. *Plant Physiol*. 21:971–978.

Torrey, J.G. 1976. Root hormones and plant growth. *Ann. Rev. Plant Phys*. 27:435–459.

Waisel, Y., and S.W. Brekle. 1987. Differences in responses of various radish roots to salinity. *Plant Soil* 104:191–194.

Waisel, Y., A. Eshel, and U. Kafkafi. 1991. *Plant Roots: The Hidden Half*. Marcel Dekker, Wageningen, The Netherlands. (*In press*.)

Walker, J.M. 1969. One-degree increments in soil temperatures affect maize seedling behavior. *Soil Sci. Soc. Am. Proc*. 33:729–736.

Wilsie, C.P. 1962. *Crop Adaptation and Distribution*. W.H. Freeman and Company, San Francisco, 448 pp.

Yang, S.F. 1980. Regulation of ethylene biosynthesis. *Hort. Sci*. 15:238–243.

Ycas, J.W., and R.W. Zobel. 1983. The response of maize radicle orientation to soil solution and soil atmosphere. *Plant Soil* 70:27–35.

Yurgalevitch, C.M. and H.W. Janes. 1988. Carbon dioxide enrichment of the root zone of tomato seedlings. *Hort. Sci.* 63:265–270.

Zobel, M.P. 1971. *Tomato Information* University of California Extension Service, Yolo County, Woodland, (*various issues—fall 1971*)

Zobel, R.W. 1972. *Genetics and physiology of two root mutants in tomato, Lycopersicon esculentum* MILL. Dissertation, Department of Vegetable Crops University of California, Davis.

Zobel, R.W. 1975. The genetics of root development. *In* J.G. Torrey and D.F. Clarkson (Eds.). *The Development and Function of Roots.* Academic Press, London, pp. 261–275.

Zobel, R.W. 1980. Rhizogenetics of soybean. *In* F.T. Corbin (Ed.). *Proceedings World Soybean Congress II.* Westview Press, Granada, pp. 73–88.

Zobel, R.W. 1986. Rhizogenetics (root genetics) of vegetable crops. *Hort. Sci.* 21:956–959.

Zobel, R.W. 1989. Steady-state control and investigation of root system morphology. In J.G. Torrey and L. Winship (Eds.). *Steady-state and Continuous Control of Root Growth.* Kluwer Academic Publishers, Dordrecht, The Netherlands, pp. 165–182.

Zobel, R.W. 1991a. Root growth and development. In: The Rhizogphere and Plant Growth. Ed: D.L. Keister and P.E. Kregan. pp. 61–71. Kluwer Academic Pub. Dordrecht, The Netherlands.

Zobel, R.W. 1991b. Genetic control of root systems. *In* Y. Waisel, A. Eshel, and U. Kafkafi (Eds.). *Plant Roots: The Hidden Half* pp. 27–38. Marcel Dekker, Wageningen, The Netherlands.

Zobel, R.W. 1990. A powerful statistical model for understanding genotype-by-environment interaction. *In* M.S. Kang (Ed.). *Genotype-By-Environment Interaction and Plant Breeding.* Louisiana State University, Baton Rouge, pp. 126–140.

Zobel, R.W., and L.W. Roberts. 1978. Effects of low concentrations of ethylene on cell division and cytodifferentiation in lettuce pith explants. *Can. J. Bot.* 56:987–990.

Zobel, R.W., M.W. Wright, and H.G. Gauch. 1988. Statistical analysis of a yield trial. *Agron. J.* 80:388–393.

Root Function in Water Transport

J.M. Baker, J.M. Wraith and F.N. Dalton

I. Introduction 53
II. Water Entry into Roots 54
III. Root Function During Drought 63
IV. Concluding Remarks 66
References 67

I. Introduction

Water is the primary constituent of plant tissue and the solvent in which plant biochemical reactions take place, but the amount required for constitutive purposes is small compared to that which passes through the plant from soil to atmosphere, serving the dual purposes of nutrient transport and evaporative cooling. The water needed to maintain foliage temperatures within a biologically comfortable range may exceed 10 kilograms per square meter of land area on sunny summer days. Acquisition of such quantities of water imposes tremendous demands on plant root systems, and the failure of roots to supply those demands is one of the principal constraints on plant productivity in many parts of the world. Consequently, root function in water transport has been an active area of research in agronomy and plant physiology, dating from the infancy of those sciences.

The work of Dixon (1914) may logically be considered as a beginning point in the modern study of roots and plant-water relations. His "cohesion theory" provided a plausible explanation of the altogether remarkable means by which water moves from the soil through plants to

1992 by Springer-Verlag New York Inc.
Advances in Soil Science, Volume 19

the atmosphere against a gravitational potential difference that can exceed 100 m in the case of large trees.

This report is not a historical treatment nor is it meant to be comprehensive in its scope. Emphasis will be placed on those aspects of root-water relations that are currently considered to be at issue. It is in some ways surprising how much is unknown or poorly understood about roots and their function in water transport, but much of that can be traced to the difficulties inherent in the study of roots. Root research in the field is tedious, laborious work that yields clear conclusions seldom and grudgingly, and it is difficult to observe root functions without affecting them in the process. Laboratory, growth chamber, and greenhouse research, though they permit the application of elegant measurement techniques, have their own hazards, primarily associated with the introduction of unwanted experimental influences traceable to unnatural rooting media and growing conditions. These can lead to conclusions that do not apply to field-grown plants or to exaggerations of the importance of observed phenomena. Thus, there is still uncertainty about many of the aspects most basic to root function in water transport.

II. Water Entry into Roots

A. Pathways

"In its passage from the soil to the atmosphere, water of the transpiration stream must traverse two cellular barriers: that between the soil and the xylem elements of the root, conveniently designated the root cortex, and that between the xylem of the leaf vein and the transpiring cell, designated leaf mesophyll. Transpirational loss of water from the mesophyll cell causes a gradient of water potential through the leaf tissue in response to which water moves from the vein. This results in a reduction of hydrostatic pressure in the xylem in response to which water moves across the root cortex from the soil." This statement by Weatherly (1963) can be considered to be a generally accepted view of the transpiration process. The route taken by water moving in the transpiration stream through plant roots has been the subject of considerable study without yet yielding to unequivocal agreement. The likely alternatives are **apoplasmic** flow, in which water travels mainly extracellularly through the "free space" in the walls between cells, and **symplasmic** flow, in which water travels intercellularly through the cytoplasm via plasmodesmata. Symplasmic flow is sometimes placed in the more general category of cell-to-cell transport, which also includes water moving across the cells and their walls, although the latter is not considered to be important to transpirational flow because of the additional resistances encountered. Water which crosses the root cortex apoplasmically must still pass at least two membranes before arriving in the xylem, due to a nearly impermeable anatomical feature known as the Casparian Strip, a deposition of wax-like compounds found at the endodermis in nearly all plants. Further complicating the issue is the fact that Casparian Strip formation begins some distance behind the root tip. However, xylem differentiation and development also lags some distance behind the region of meristematic activity. Thus, while it is expected that radial resistance to flow is lower in new root segments than in more mature tissue, this may be counterbalanced by higher axial resistance. Much of the evidence cited in favor of apoplasmic flow has been in the form of dye studies which show extracellular movement up to the endodermis and movement within cells from there into the stele. However, such results may be experimental artifacts to the extent that cell membranes reject passage of the tracers used. Arguments against apoplasmic flow have suggested that reported permeabilities of the extracellular "free space" are too low to support observed flow rates at given potential gradients (Newman, 1976), possibly due to the hydrophobic nature of the cell wall compounds (Passioura, 1988).

Descriptions put forth for symplasmic flow posit direct hydraulic connections between the cytoplasm of contiguous cells, and, while such structures are clearly present, their internal structure and permeability to water transport remain unclear. Recent work by Steudle and Jeschke (1983), Jones et al. (1984), and Steudle (1989) has addressed questions about flow paths by comparing the permeabilities of individual cells with those of intact roots. The premise for this is that if whole root permeabilities are of a similar magnitude to those of individual cells, then a pathway passing through only one or two cells (apoplasmic) is indicated, while a whole root permeability many times greater than the individual cell permeability suggests a symplasmic route. The evidence varies among species, which may well explain some of the historical disagreement and experimental discrepancies with respect to pathways. These results show, for example, that, in barley (*Hordeum vulgare*), the whole root conductances are smaller by an order of magnitude than those of the individual cells, indicating symplasmic transport. On the other hand, in maize, (*Zea mays*), the cell and whole root conductances are of similar magnitude, suggesting substantial apoplasmic transport. The overall evidence seems to favor symplasmic transport as the generally more important path for water influx, though the extent to which this is true must vary among species in accord with variability in the relative resistances to flow of the different pathways.

B. Coupling of Water and Solute Uptake

Any quantitative discussion of resistance to water flow must always be referenced to some specific mathematical model of the transport process under consideration. For water flow in the soil-plant-air continuum

(SPAC), the most prevalent and accepted mathematical description used today was proposed by van den Honert (1948). His model is analogous to the set of rules describing current flow in an electric circuit. That is, water flows through the plant at a rate directly proportional to the water potential gradient across the plant and inversely proportional to the plant resistance. This description assumes that the plant root system behaves as a passive element in the transpiration process and that the flow can be described by knowledge of soil and plant water potentials in conjunction with the appropriate tissue resistance. The relative importance of this component in the SPAC is discussed in a later section.

Dalton et al. (1975) and Fiscus (1975) used a fundamentally different approach for characterizing the flow of both water and solute through plant root tissue. They assumed that the nutrient solution is separated from the xylem solution by an effective semi-permeable membrane with active ion transport capabilities and then used the theory of irreversible thermodynamics to describe the flow characteristics of the system. This approach, which couples water and solute flow, is considered to be phenomenological in that knowledge of specific mechanisms of ion transport, and the exact location of barriers to water or ion transport is not necessary. The analysis was applied to plant root systems and the resulting predictive capabilities were compared with experimental observations. The effective semi-permeable membrane is assumed to operate within the system of cells separating the nutrient solution from the xylem solution and consists of the epidermis, cortex, endodermis, and stele.

Emphasis has been put on the endodermis as representing a barrier to solute flow. Hendricks (1966) and Oberlander (1966) reviewed the origin of this concept and showed that it is largely based on morphological considerations. However, evidence that a barrier exists has more than a morphological foundation. Bernstein and Gardner (1961) analyzed the diffusion data of Smith (1960) to show that a barrier to solute flow exists. Ginsburg and Ginzburg (1970) showed that a barrier exists in a root from which the stele and the inner tangential wall of the endodermal cells have been removed. While the epidermis is sometimes implicated as a solute barrier, Epstein (1960) summarized evidence that the free space of the cortex is in virtually instant communication with the nutrient solution. The only true consensus is that a barrier does exist somewhere.

If a semi-permeable membrane separates the soil solution from the xylem solution, the transpiration process will depend not only on a hydraulic gradient and associated flow resistance, but also on an osmotic component whose effectiveness depends on the quality of the semi-permeable membrane (reflection coefficient), the diffusive resistance of the membrane (osmotic permeability coefficient), and the rate of metabolically driven ion uptake (active transport). This formulation yields a relationship between water flux and hydraulic gradient that is more complicated because plant resistance now includes all manifestations of the osmotic and active ion transport variables. More importantly, this formulation predicts the observed decrease in xylem ion concentration with increased transpiration rate; simulates the effect of active ion uptake on the hydraulic pressure gradient required for a given transpiration rate; shows the effect of nutrient solution ion concentration, active ion uptake, and transpiration rate on ion flux into the plant; and exhibits a nonlinear relation between xylem hydraulic pressure and flow through the roots. This formulation also predicts a significant temperature dependence of transpiration in terms of the temperature dependence of active ion transport, and it even gives a quantitative expression for root pressure which agrees with experimental observations.

To be sure, such a model is more complicated than the Ohm's law analogy and may be unduly complicated where the primary objective is to show the effects of root water uptake on the space-time distributions of soil water content. However, when the primary objective is to simulate plant response to water and salt stress, the simple Ohm's law analogy may be inadequate. It becomes essential to, at least qualitatively, understand the many relationships that exist between transpiration, hydraulic gradient, ion flux, xylem solution concentration, active ion transport, and osmotic potential of the nutrient solution. More recently, Cilentano et al. (1988) used a similar approach to simulate flow characteristics through a series arrangement of semi-permeable membranes that included chemical reactions. Others (Katou et al., 1987; Taura et al., 1988; Katou and Taura, 1989) are also using membrane transport theory to model water movement due to transpiration and ion transport at the cellular level. These concepts have also proved useful in the development of a dynamic salinity stress index (Dalton and Poss, 1989). This index is defined in terms of the relative rates of salt flux to the shoot. It incorporates, at least qualitatively, the biophysical transport properties of the root and all other parameters in the soil-plant-air continuum that affect transpiration. Far more effort, however, has been put into analyzing the flux-force relations in terms of van den Honert's Ohm's law analogy (1948), wherein the concept of plant resistance is emphasized.

C. Factors Affecting Root Permeability

There are a bewildering number of literature reports on the effects of a wide variety of physical, chemical, and cultural variables on water uptake by plants. Many are difficult to interpret due to experimental uncertainties and confounding influences, but several points are worthy of mention. As we have seen in the previous section, membrane permeability and active solute uptake are both critical components in water uptake; thus, factors affecting either component can influence water transport in roots. Among these factors are soil salinity, oxygen deficiency (Kramer, 1983), nitrogen deficiency (Radin and Boyer, 1982), metabolic poisons such as herbicides, and temperature.

Shalhevat et al. (1976) investigated the effect of salinity on root permeability for tomato (*Lycopersicon lycopersicum*) and sunflower (*Helianthus annuus* L.) and found no appreciable influence. Because the relative permeability of roots to water and ions is considered a significant factor relating to plant salt tolerance, Dalton and Poss (1989) measured the effects of root temperature and osmotic stress on water use, ion transport, and yield. Tomatoes were grown in a greenhouse at various levels of salinity and root temperature, with the shoot temperature held uniform. Yield, water use, and salt tolerance increased with increasing temperature to a maximum and then decreased. The response was similar to the temperature dependence of enzyme activity. The variability of transport parameters in the root is vitally important in relation to salt loading in the shoot. These experiments demonstrate the role that root transport properties can have on plant responses to salinity and show that we can distinguish between genetic and biophysical transport parameters determining plant salt tolerance.

The temperature sensitivity of root uptake can be of considerable importance under certain circumstances, as in climates where high evaporative demand may occur early in the growing season when soils may still be relatively cool. Dalton and Gardner (1978) discussed the temperature dependence of water uptake from a theoretical standpoint, considering temperature effects on viscosity, solute diffusivity, and rate constants for active solute uptake. They derived a general equation for the effect of temperature on uptake and explained the observed temperature dependence of water transport in terms of the temperature dependence of metabolic-driven ion transport.

Markhart et al. (1979) showed the dramatic effect of low temperature on conductance in both soybean and broccoli, and they were able to show that the effect was indeed a decrease in root conductance rather than a change in the osmotic driving force. Arrhenius plots of flow rate versus the reciprocal of temperature for soybean (*Glycine max* L.) showed clear breaks consistent with those associated with sudden phase changes in membrane lipids; further, the data showed that plants could acclimate, with some evidence that acclimation was associated with a decrease in saturation of fatty acids in the roots. Plants grown at lower temperatures exhibited breaks at lower temperatures, although their activation energies above the break temperature were not different. Broccoli (*Brassica oleraceae* L.) was different in two respects: there were no breaks in the Arrhenius plots and there were different activation energies for the warm- and cold-grown plants. That is, the conductance of the cold-grown plants was less sensitive to temperature than that of the warm-grown plants.

Several papers have reported membrane changes associated with drought stress. Liljenberg and Kates (1985) showed that the lipid/protein ratio of oat roots decreased with increasing drought stress and that lipid composition changed as well, with the degree of unsaturation decreasing as drought stress intensified. They did not measure associated changes in conductance, but others have noted decreased conductance in droughted roots (Kramer, 1950; Blizzard and Boyer, 1980).

D. The Relative Importance of Plant and Soil Resistances

The relative significance of plant and soil resistances has been a topic of debate for many years, dating at least to Gardner's (1960) analytical solution for flow to roots. In that and a subsequent manuscript (Gardner and Ehlig, 1962), it was suggested that the soil might provide a significant impediment to flow at water potentials much higher than the commonly accepted "wilting point" of $-1.5\,\text{MPa}$. Newman (1969a,b) subsequently criticized these and several other manuscripts with similar conclusions and made a theoretical case for the argument that soil resistance should be small relative to that of the plant until soils are near the $-1.5\,\text{MPa}$ range, assuming typical rooting densities. He admitted that direct evidence in support of his position was lacking due to experimental difficulties which have hindered resolution of this argument to the present day.

Blizzard and Boyer (1980) present one of the few cases in which water potentials were measured in the soil, in the root system, and in the leaves at a number of different transpiration rates. The root water potential measurement was made isopiestically on secondary root tissue of soybean (*Glycine max* L.). The authors used these data to partition resistance to flow between a soil resistance, which included resistance in the outer root tissue as well, and a plant resistance, which they defined to include resistance across the endodermis plus xylem resistance between the roots and foliage. They showed that, as the soil dried, both the soil and plant resistances increased, with the ratio of plant to soil resistance decreasing from a value somewhat greater than 2 in wet soil to approximately 1.5 in dry soil. This is noteworthy in several respects. First, the increase in plant resistance with soil drying was contrary to the assumptions of constant plant resistance made by Gardner and Ehlig (1962), Cowan (1965), and others. Blizzard and Boyer speculated that this increase might be due to some combination of cavitation, about which more will be said later, and increased root resistance, perhaps due to suberization. Secondly, the soil resistance was of the same magnitude as the plant resistance over the entire range, even in wet soil, which is contrary to the assumptions made by Newman (1969a) and many others. This is all the more remarkable in light of the fact that Blizzard and Boyer were working with potted plants which had root densities of nearly $10\,\text{cm}\,\text{cm}^{-3}$, quite high for soybean. However, it must be noted that they included in their soil resistance an indeterminate portion of the outer root resistance, which could be substantial if the flow in soybean is primarily symplasmic. Blizzard and Boyer interpreted their results as generally supportive of Newman's (1969a) assertions about the relative importance of plant versus soil resistance.

In a more recent theoretical exercise, Hulugalle and Willatt (1983) have raised anew the possibility that soil resistance may be important even at relatively high water potentials. They suggested that root densities in the field, particularly in certain soil layers, may be much lower than the densities quoted by Newman (1969a), and they made the further point that, in structured soils, roots tend to clump in large pores and along ped faces, meaning that effective root densities are much lower than those obtained by assuming uniform root distribution. Passioura (1988) has made this point as well. Further, as Caldwell (1976) noted, there is evidence that root permeability may vary considerably with position along a root, which could lead to local zones of depletion in which resistance to flow might reach appreciable levels even as the bulk soil potential remained quite high. Hulugalle and Willatt (1983) made the additional point that soils vary tremendously in their conductivity-potential relations, so that there should be considerable variability in soil resistance among soil types at any given water potential. They proposed, as had Reicosky and Ritchie (1976), that soil hydraulic conductivity provides a better indicator than water potential of the point at which soil resistance becomes limiting. Their calculations suggest that soil resistance becomes appreciable when soil conductivities reach 10^{-8} to $10^{-9}\,\mathrm{m\,d^{-1}}$ for plants with rooting densities in the range of $1\,\mathrm{cm\,cm^{-3}}$. This is similar to the conclusions reached by Reicosky and Ritchie (1976), who found that these conductivities corresponded to a range of water potentials of -0.1 to $-0.8\,\mathrm{MPa}$ for the soils they used. To the extent that root distributions deviate from uniform, soil resistance will become appreciable at even higher water potentials.

E. Reverse Flow

Reverse flow refers to the movement of water from roots into soil. Gradients favoring such movement can occur when part of a plant's root system is in wet soil and part of it is in dry soil, if evaporative demand is sufficiently low that the dry soil represents a stronger sink for water than the foliage. Rosene (1944) showed many years ago with potometer experiments that water flow into the roots could be reversed if the osmotic potential of the external solution was lowered, and Jensen et al. (1961) showed that resistance to flow through a whole plant was independent of the direction of flow. However, reverse flow was apparently never considered to be of much importance, probably because it was felt that gradients favoring such movement were rare. The advent of numerical simulation brought renewed interest in the question when it was observed that finite difference models often predicted substantial transport of water through the roots from wet soil layers to drier layers during periods of low evaporative demand (van Bavel and Ahmed, 1976). A number of authors have arbitrarily prohibited such flow in their models with numerical "rectifiers," presumably because they doubted that it would actually occur. This view has been supported in part by several studies in which little reverse flow occurred, despite favorable gradients. These include work by Molz and Peterson (1976) with cotton, Nobel and Sanderson (1984) with desert succulents, and Dirksen and Raats (1985) with alfalfa. All three reported substantially higher resistances to reverse flow than to normal uptake.

By contrast, a number of reports of measurable quantities of water lost from roots to dry soil can be found, including work by McWilliam and Kramer (1968) with Mediterranean grasses (*Phalaris tuberosa*), Mooney et al. (1980) with mesquite (*Prosopis tamarugo*), Baker and van Bavel with Bermudagrass (*Cynodon dactylon*) (1986) and cotton (*Gossypium hirsutum*) (1988), and Richards and Caldwell (1987) and Caldwell and Richards (1989) with sagebrush (*Artemisia tridentata*). Baker and van Bavel (1986) found that the resistance to reverse flow in Bermudagrass was only slightly higher than the resistance to uptake. With cotton, resistances were also similar but no outflow was observed until soil water potentials dropped below −1.0 MPa. This was apparently due to large offsets in the flow versus potential gradient relation, which have also been observed by others in studies of cotton uptake.

The work of Richards and Caldwell (1987) and Caldwell and Richards (1989) provides an elegant demonstration of the process of reverse flow. They measured soil water potentials automatically and continuously in the upper layers of the root zone of *Artemisia* and detected consistent overnight increases far too large to explain with Darcian flow. With artificial lights and gas exchange measurements of transpiration, they were able to show that the overnight increases could be entirely prevented if evaporative demand were maintained during the night, and, conversely, that the overnight increases would continue into the next day if transpiration were suppressed by covering the foliage. They postulated that negative flow might be beneficial to arid rangeland species with sparse, but deep, root systems by allowing them to take up water around the clock from deeper, wetter soil layers. The plants would thus be banking the water taken up at night in shallow dry soil layers where it could help the plant meet evaporative demand during the following day. Baker and van Bavel (1988) suggested that a similar mechanism might allow more efficient application of water in alternate furrow irrigation.

Passioura (1988) has offered some possible explanations for the differing experimental results with respect to reverse flow. The most obvious is that species and cultivars differ in their resistance to outflow. One reason for such differences might be the diurnal fluctuations in root hydraulic resistance reported for some species by Hagan (1949) and Parsons and Kramer (1974), since higher nocturnal resistance would tend to minimize reverse flow. A second possibility mentioned by Passioura (1988) is that outflow causes accumulation of solute in the xylem of some species, which tends to reduce the gradient in potential. Whether this is likely, given the low solute concentrations typically found in xylem sap, is a question best addressed by the coupled water/solute transport approach of Dalton et al. (1975) and Fiscus (1975) discussed earlier.

F. Vapor Phase Absorption

Regardless of the model used to describe root water uptake, there appears to be a general consensus that all water transfer from the soil to the root is in the liquid phase. For soil, it is known that the vapor phase component is usually a small fraction of the total water transport. However, root shrinkage and soil water depletion due to transpiration and drainage may create vapor spaces in portions of the root zone, causing the root surface to lose contact with the liquid phase water (Huck et al., 1970; Herkelrath et al., 1977). Even in this situation, it is tacitly assumed that water is extracted in the liquid phase. From time to time, however, it has been argued from circumstantial evidence that there may be appreciable vapor phase water absorption by plant roots (Philip, 1957; Bonner, 1959; Cowan and Milthorpe, 1968). Others have argued sensibly against this notion (Bernstein et al., 1959). However, this hypothesis was not directly testable until recently when Dalton (1989) used the properties of stable isotopes of hydrogen in the water molecule to distinguish between liquid and vapor phase water absorption by plant roots. The vapor pressure of the lighter H_2O molecules is slightly greater than the naturally occurring, heavier HDO molecules (D = deuterium atom). Therefore, vapor phase water in equilibrium with liquid phase water will have a slightly lower ratio (HDO)/(H_2O) than that in the liquid phase. A vapor component to the water extraction process will manifest itself by depleting a larger fraction of the lighter H_2O molecule than the heavier HDO molecule from the residual soil water. In these experiments, a vapor phase component was indicated by the observed increase in isotopic abundance of HDO with increasing depletion of soil water. That is, water depletion in the root zone was accompanied by a distillation process (vapor transport) which concentrates the heavier HDO water molecules in the soil solution. If all root water absorption was in the liquid phase, there would be no distillation and the stable isotope concentration would have remained constant with soil water depletion. If vapor absorption is assumed to be described by a Rayleigh distillation process, then the observed rate of increase in isotopic abundance of the residual soil water indicates a very small component of vapor absorption. While other distillation processes are possible, quantitative estimates of vapor flux require further work. In any case, this method offers a direct experimental method for studying a previously unverifiable assumption concerning the mechanism for water absorption by plant roots.

G. Influence of VAM Fungi

Vesicular arbuscular mycorrhizal (VAM) fungi are common in many vegetation types in nearly all climatic zones (Fitter, 1989). These fungi form symbiotic associations by infecting plant root systems and sharing soil resources in exchange for a ready supply of carbon from the host plant. Mycelial hyphae may extend well away from the root surfaces into the bulk soil, significantly increasing the plant's sphere of access to relatively immobile nutrients, particularly phosphorus. These associations have also been reported to enhance water uptake, transpiration, and photosynthesis of host plants under certain conditions (Allen et al., 1981; Dixon et al., 1983; Walker et al., 1989), although accounts describing no positive effects are equally common (Levy et al., 1983; Graham et al., 1986; Lehto, 1989). Little conclusive evidence exists of a direct influence on root water uptake and transport. In most cases, improved plant water relations may be explained by the indirect effects of enhanced phosphorus nutritional status (Graham, 1987; Puppi and Bras, 1989). Severely limited rooting volumes employed in many laboratory investigations make an extension of results to field conditions questionable, further hindering resolution of these issues. Under typical agricultural conditions of adequate phosphorus supply for plant growth and development, it is unclear whether VAM fungi will have any significant beneficial influence on root water uptake and transport.

III. Root Function During Drought

A. Cavitation Repair

There is growing evidence that xylem cavitation (blockage of xylem elements by formation of gas bubbles) may be one of the primary deleterious effects of drought stress in plants. Tyree and Sperry (1989) have written a comprehensive review that details how such emboli form, how their formation can be detected, what the consequences of cavitation are, and how plants repair the damage that is caused. Cavitation occurs in response to both freezing and decreases in water potential, but it is the latter that is of primary importance in herbaceous plants.

Milburn and Johnson (1966) first detected cavitation events by listening to the audible "pops" produced in the process. Tyree and Dixon (1983) later showed that the measurement could be made more easily and automatically in the ultrasonic range. Tyree et al. (1986) used the ultrasonic method in field-grown maize and found that a substantial number of cavitation events occurred on a daily basis when xylem water potentials dropped below −1.0 MPa, even under irrigated conditions. In unirrigated maize (*Zea mays*), where the xylem potentials reached lower values, they estimated that half of the xylem vessels cavitated on a daily basis. This substantially reduced the hydraulic conductance of the plants, though in both irrigated and unirrigated maize they found that most cavitation damage was repaired overnight.

Repair of cavitation is thought to be facilitated by root pressures, i.e., positive or near positive pressures developed in the xylem at night due to solute uptake. Presumably, it becomes more and more difficult to generate the root pressures required to repair emboli as soil water potentials decrease. It is plausible that both the vulnerability of plants to cavitation and their ability to generate the root pressures necessary for remediation vary among species and cultivars and thus may be a source of differences in drought tolerance.

B. Root Osmotic Adjustment

Osmotic adjustment is a decrease in cell osmotic potential due to increased numbers of solute molecules rather than decreased cell volume (Turner and Jones, 1980). Increased solute concentration as a result of osmotic adjustment aids in the maintenance of favorable cell turgor pressures as total cell water potential decreases. In non-halophytic plants, this phenomenon appears to be a passive response to drought, apparently the result of cell division and expansion being more sensitive to water stress than photosynthesis, causing organic photosynthate ions to accumulate (Munns, 1988). Differences among species in their capacity for osmotic adjustment are likely a result of differences in the degree to which growth is constrained relative to photosynthesis.

Osmotic adjustment in leaves has been observed in a number of species, but it is more difficult to document for roots. Turner et al. (1987) found that osmotic adjustment of roots was negligible compared to that of leaves in three of four lupin (*Lupinus* sp.) species they examined. However, Oosterhuis and Wullschleger (1987) noted that, although cotton leaves exhibited larger absolute magnitudes of osmotic adjustment, osmotic adjustment of roots was much greater when expressed on a percentage basis. Parker and Pallardy (1988) observed osmotic adjustment in seedling root tissue in only one of three *Quercus* species investigated while the leaves of all three species adjusted osmotically. They noted, however, that the root tissue osmotic potential at the turgor loss point in the two other species was lower in response to drought stress as compared to control conditions because of increased cell wall elasticity. While this is not osmotic adjustment in the strict sense, it also serves to maintain turgor in the face of diminishing soil water availability. In fact, because many studies of root and shoot responses to drought do not determine the osmotic potential at full turgor (Sharp and Davies, 1979; McGowan et al., 1984; Westgate and Boyer, 1985), it is impossible to ascertain whether lowered potentials are indicative of true osmotic adjustment or increased cell wall elasticity or both.

Because the accumulation of solute molecules presumably comes at the expense of plant growth, osmotic adjustment is probably of value primarily as a survival mechanism. Maintenance of root turgor may allow continued growth into more favorable root zones and aid in the re-filling of cavitated xylem elements. Perhaps most importantly, maintenance of some degree of cell turgor in root apices may preserve these meristematic regions in dry soil (i.e., near the soil surface), permitting growth and uptake to resume when soil conditions become more favorable.

C. Root to Shoot Communication

The conventional view of the effect of drought on transpiration rate has been that it is primarily hydraulic in nature, (i.e., declining soil water potential causes a concomitant decrease in bulk leaf water potential), eventually resulting in the loss of guard cell turgor and, hence, a reduction in stomatal aperture. This has recently come under critical review. Increasing evidence of direct root to shoot communication of physical conditions in the rhizosphere indicates that the roots may have a direct role in regulating plant water transport independent of their hydraulic linkage to shoot water status. Several researchers have demonstrated a decline in stomatal conductance as soil water was depleted even though leaf water status was not measurably affected (Bates and Hall, 1981; Blackman and Davies, 1985; Gollan et al., 1985; Turner et al., 1985; Kuppers et al., 1988). Gollan et al. (1986) placed pots containing wheat plants (*Triticum aestivum*) in a pressure chamber apparatus and, by increasing the pressure on the enclosed root zone, were able to maintain leaf turgor at nearly constant levels as the soil dehydrated. They found that stomatal conductance decreased sharply as soil water content fell below a critical range, regardless of whether the roots were or were not pressurized. These observations are consistent with a hypothesis of roots communicating information regarding soil physical conditions directly to the shoot by some mechanism other than the energy status of water within the xylem. Because soil strength usually increases upon dehydration, it is not clear whether the roots may be sensing soil water status, increased mechanical resistance, or both.

Good circumstantial evidence has been found that this response is mediated by some chemical signal from the root tips, although the exact mechanism(s) remains controversial. Abscisic acid (ABA) has been shown to be a powerful promoter of stomatal closure (Jones and Mansfield, 1970). It has been suggested that growing root tips continually produce cytokinins which are transported to the leaves via the xylem, where they counteract the inhibitory effects of native ABA (Blackman and Davies, 1985; Davies et al., 1986). When the root tips become dehydrated, according to this hypothesis, the supply of cytokinins is interrupted and stomatal closure begins. Synthesis of ABA by water-stressed root tips and its transport to leaves in the transpiration stream has also been postulated. Several papers have reported an increase in ABA production and/or content in water-stressed roots (Lachno and Baker, 1986; Zhang and Davies, 1987, 1989, 1990; Zhang et al., 1987), and Zhang and Davies (1987, 1989, 1990) provide evidence that this ABA may indeed be transported to the leaves. Munns and King (1988), however, found that transpiration rates of detached wheat leaves were reduced by xylem sap even after the ABA was removed.

The possibility of experimental artifact due to the unnatural plant/soil conditions imposed during many of these investigations is difficult to overlook. Berliner and Oosterhuis (1987) compared the onset of crop water stress for spring wheat growing in 1.0 m deep field lysimeters and in the surrounding field. They found that water stress developed gradually in the field-grown plants, with no identifiable threshold value for soil water content. However, the onset of water stress was relatively abrupt for plants within the lysimeters; stomatal conductance and leaf water potential responses to soil water status were strikingly similar to those reported in studies cited above where plants were grown in pots. They attributed these results to dissimilar distributions of soil water and roots as measured within the confining lysimeter compared to the field situation. Saab and Sharp (1989) divided the root systems of several maize plants into two regions, one of which was well watered and one which was allowed to dry. They found that leaf elongation was inhibited compared to the control plants (both root zones were kept well watered) under conditions of relatively rapid soil drying, but not when the soil water was depleted more slowly. Careful field experimentation should indicate whether non-hydraulic signals from plant roots are an important component of whole-plant/water relations under typical growing conditions and what influence they may have on seasonal plant growth.

IV. Concluding Remarks

The importance of the subject of root function in water transport is obvious, given increasing constraints on agricultural water use in many parts of the world and the likelihood of increasing evaporative demand associated with global climate change. Though the capabilities of genetic engineering are not yet clear, questions are already asked about the desirability of different root traits with respect to plant /water relations. At the current level of understanding, biophysicists and physiologists are unable to provide the necessary answers, despite many years of research. The body of knowledge on the subject is in some ways a catalog of isolated observations, and a coherent picture of root function in situ is lacking.

It is evident that root systems are not simply "soda straws," transporting water in response to external pressures. Irreversible thermodynamics provides a framework for examining the way in which water flow is linked with solute uptake and hence metabolism, and the intriguing observations of root influences on shoot behavior suggest a direct means by which roots may exert control over the rate of water use by plants. These areas should continue to generate interest and research, but other areas are in need of attention as well. More work is needed on the effects of temperature on uptake and on the recovery of function following chilling. The aftereffects of drought, including the recovery of absorptive function following the rewetting of desiccated or suberized roots, are also subjects of considerable importance. Finally, the deceptively simple question "How much root system does a plant need?" requires a better understanding of the variations in root permeability as a function of time and position and the effect of soil structure and chemistry on the distribution of roots. In all of these areas, progress depends on the development of innovative methods and instrumentation that will allow the study of root function with minimal interference and under realistic growth conditions.

References

Allen, M.F., W.K. Smith, T.S. Moore, and M. Christensen. 1981. Comparative water relations and photosynthesis of mycorrhizal and non-mycorrhizal *Bouteloua gracilis* H.B.K. Lag ex Steud. *New Phytol.* 88:683–693.

Baker, J.M., and C.H.M. van Bavel. 1986. Resistance of plant roots to water loss. *Agron. J.* 78:641–644.

Baker, J.M., and C.H.M. van Bavel. 1988. Water transfer through cotton plants connecting soil regions of differing water potential. *Agron. J.* 80:993–997.

Bates, L.M., and A.E. Hall. 1981. Stomatal closure with soil water depletion not associated with changes in bulk leaf water status. *Oecologia* (Berlin) 50:62–65.

Berliner, P.R., and D.M. Oosterhuis. 1987. Effect of root and water distribution in lysimeters and in the field on the onset of crop water stress. *Irrig. Sci.* 8:245–255.

Bernstein, L., and W.R. Gardner. 1961. Perspective on function of free space in ion uptake by roots. *Science* 133:1482–1483.

Bernstein, L., W.R. Gardner, and L.A. Richards. 1959. Letters; Is there a vapor gap around plant roots? *Science* 129:1750–1753.

Blackman, P.G., and W.J. Davies. 1985. Root to shoot communication in maize plants of the effects of soil drying. *J. Exp. Bot.* 36:39–48.

Blizzard, W.E., and J.S. Boyer. 1980. Comparative resistance of the soil and the plant to water transport. *Plant Physiol.* 66:809–814.

Bonner, J. 1959. Water transport. *Science* 129:447–459.

Caldwell, M.M. 1976. Root extension and water absorption. *In* O.L. Lange, L. Kappen, and E.-D. Schulze (Eds.). *Ecological Studies, Analysis and Synthesis, Vol. 19. Water and Plant Life*. Springer-Verlag, Berlin, pp. 63–85.

Caldwell, M.M., and J.H. Richards. 1989. Hydraulic lift: Water efflux from upper roots improves effectiveness of water uptake by deep roots. *Oecologia* (Berlin) 79:1–5.

Cilentano, F.C., G. Monticelli, and F. Cottini. 1988. A simulation of mass transport across series arrays of membranes with chemical-reaction-coupled solute flow. *Trans. Soc. Comp. Sim.* 5:43–70.

Cowan, I.R. 1965. Transport of water in the soil-plant-atmosphere system. *J. Appl. Ecol.* 2:221–239.

Cowan, I.R., and F.L. Milthorpe. 1968. Plant factors influencing the water status of plant tisues. *In* T.T. Kozlowski (Ed.). *Water Deficits and Plant Growth*. Academic Press, New York, pp. 137–193.

Dalton, F.N. 1989. Plant root water extraction studies using stable isotopes. *In* B.C. Loughman, O. Gasparikova, and J. Kolek (Eds.). *Structural and Functional Aspects of Transport in Roots*. Kluwer Academic Publishers, Dordrecht, the Netherlands, pp. 151–155.

Dalton, F.N., and W.R. Gardner. 1978. Temperature dependence of water uptake by plant roots. *Agron. J.* 70:404–406.

Dalton, F.N., and J.A. Poss. 1989. Water transport and salt loading: A unified concept of plant response to salinity. *Int. Soc. For. Hort. Sci. V. Symposium, Scheduling Of Irrigation For Vegetable Crops Under Field Condition*. Maratea, Italy, p. 27.

Dalton, F.N., P.A.C. Raats, and W.R. Gardner. 1975. Simultaneous uptake of water and solutes by plant roots. *Agron. J.* 67:334–339.

Davies, W.J., J. Metcalfe, T.A. Lodge, and A.R. da Costa. 1986. Plant growth substances and the regulation of growth under drought. *Aust. J. Plant Physiol.* 13:105–125.

Dirksen, C., and P.A.C. Raats. 1985. Water uptake and release by alfalfa roots. *Agron. J.* 77:621–626.

Dixon, H.H. 1914. *Transpiration and the Ascent of Sap in Plants*. MacMillan, New York.

Dixon, R.K., S.G. Pallardy, H.E. Garrett, G.S. Cox, and I.L. Sander. 1983. Comparative water relations of container-grown and bare-root ectomycorrhizal and nonmycorrhizal *Quercus velutina* seedlings. *Can. J. Bot.* 61:1559–1565.

Epstein, E. 1960. Spaces, barriers, and ion carriers: Ion absorption by plants. *Am. J. Bot.* 49:393–399.

Fiscus, E.L. 1975. The interaction between osmotic- and pressure-induced water flow in plant roots. *Plant Physiol.* 55:917–922.

Fitter, A.H. 1989. The role and ecological significance of vesicular-arbuscular mycorrhizas in temperate ecosystems. *Agric. Ecosystems Environ.* 29:137–151.

Gardner, W.R. 1960. Dynamic aspects of water availability to roots. *Soil Sci.* 89:63–73.

Gardner, W.R., and C.F. Ehlig. 1962. Some observations on the movement of water to plant roots. *Agron. J.* 54:453–456.

Ginsburg, H., and B.A. Ginzburg. 1977. Radial water and solute flow in roots of *Zea mays*. II. Ion fluxes across root cortex. *J. Exp. Bot.* 21:593–604.

Gollan, T., J.B. Passioura, and R. Munns. 1986. Soil water status affects the stomatal conductance of fully turgid wheat and sunflower leaves. *Aust. J. Plant Physiol.* 13:459–464.

Gollan, T., N.C. Turner, and E.-D. Schulze. 1985. The responses of stomata and leaf gas exchange to vapor pressure deficits and soil water content. III. In the sclerophyllous woody species *Nerium oleander*. *Oecologia* (Berlin) 65:356–362.

Graham, J.H. 1987. Non-nutritional benefits of VAM fungi-do they exist? *In* D.M. Sylvia, L.L. Hung, and J.H. Graham (Eds.). *Mycorrhizae in the Next Decade: Practical Applications and Research Priorities*. Proc. of the 7th North American Conference on Mycorrhizae, Gainesville, Florida, May 3–8, 1987, pp. 237–239.

Graham, J.H., J.P. Syvertsen, and M.L. Smith, Jr. 1986. Water relations of mycorrhizal and phosphorus-fertilized non-mycorrhizal citrus under drought stress. *New Phytol.* 105:411–419.

Hagan, R.M. 1949. Autonomic diurnal cycles in the water relations of nonexuding detopped root systems. *Plant Physiol.* 24:441–454.

Hendricks, S.B. 1966. Salt entry into plants. *Soil Sci. Soc. Am. Proc.* 30:1–7.

Herkelrath W.N., E.E. Miller, and W.R. Gardner. 1977. Water uptake by plants. II. The root contact model. *Soil Sci. Soc. Am. J.* 41:1039–1043.

Huck, M.G., Klepper, B., and H.M. Taylor. 1970. Diurnal variations in root diameter. *Plant Physiol.* 45:529–530.

Hulugalle, N.R., and S.T. Willatt. 1983. The role of soil resistance in determining water uptake by plant root systems. *Aust. J. Soil Res.* 21:571–574.

Jensen, R.D., S.A. Taylor, and H.H. Wiebe. 1961. Negative transport and resistance to water flow through plants. *Plant Physiol.* 36:633–638.

Jones, H., A.D. Tomos, R.A. Leigh, and R.G. Wyn Jones. 1984. Water-relation parameters of epidermal and cortical cells in the primary root of *Triticum aestivum* L. *Planta* 158:230–236.

Jones, R.J., and T.A. Mansfield. 1970. Suppression of stomatal opening in leaves treated with abscisic acid. *J. Exp. Bot.* 21:714–719.

Katou K., and T. Taura. 1989. Mechanism of pressure-induced water flow across plant roots. *Protoplasma* 150:124–132.

Katou, K., T. Taura, and M. Furumoto. 1987. A model for water transport in the stele of plant roots. *Protoplasma* 140:123–132.

Kramer, P.J. 1950. Effects of wilting on the subsequent intake of water by plants. *Am. J. Bot.* 37:280–284.

Kramer, P.J. 1983. *Water Relations of Plants*. Academic Press, Orlando, Florida, 489 pp.

Kuppers, B.I.L., M. Kuppers, and E.-D. Schulze. 1988. Soil drying and its effect on leaf conductance and CO_2 assimilation of *Vigna unguiculata* (L.) Walp. I. The response to climatic factors and to the rate of soil drying in young plants. *Oecologia* (Berlin) 75:99–104.

Lachno, D.R., and D.A. Baker. 1986. Stress induction of abscisic acid in maize roots. *Physiol. Plantarum* 68:215–221.

Lehto, T.H. 1989. Effects of mycorrhiza and drought on photosynthesis and water relations of Sitka spruce. *Agric. Ecosystems Environ.* 28:299–303.

Levy, Y., J.P. Syvertsen, and S. Nemec. 1983. Effect of drought stress and vesicular arbuscular mycorrhiza on citrus transpiration and hydraulic conductivity of roots. *New Phytol.* 93:61–66.

Liljenberg, C., and M. Kates. 1985. Changes in lipid composition of oat root membranes as a function of water deficit stress. *Can. J. Biochem. Cell Biol.* 63:77–84.

Markhart, A.H., E.L. Fiscus, A.W. Naylor, and P.J. Kramer. 1979. Effect of temperature on water and ion transport in soybean and broccoli systems. *Plant Physiol.* 64:83–87.

McGowan, M., P. Blanch, P.J. Gregory, and D. Haycock. 1984. Water relations of winter wheat. 5. The root system and osmotic adjustment in relation to crop evaporation. *J. Agric. Sci. Camb.* 102:415–425.

McWilliam, J.R., and P.J. Kramer. 1968. The nature of the perennial response in mediterranean grasses. *Aust. J. Agric. Res.* 19:381–395.

Milburn, J.A., and R.P.C. Johnson. 1966. The conduction of sap II. Detection of vibrations produced by sap cavitation in *Ricinus* xylem. *Planta* 66:43–52.

Mooney, H.A., S.L. Gulmon, P.W. Rundel, and J. Ehleringer. 1980. Further observations on the water relations of *Prosopis tamarugo* of the northern Atacama desert. *Oecologia* (Berlin) 44:177–180.

Molz, F.J., and C.M. Peterson. 1976. Water transport from roots to soil. *Agron. J.* 68:901–904.

Munns, R. 1988. Why measure osmotic adjustment? *Aust. J. Plant Physiol.* 15:717–726.

Munns, R., and R.W. King. 1988. Abscisic acid is not the only stomatal inhibitor in the transpiration stream of wheat plants. *Plant Physiol.* 88:703–708.

Newman, E.I. 1969a. Resistance to water flow in soil and plant. I. Soil resistance in relation to amounts of root: Theoretical estimates. *J. Appl. Ecol.* 16:1–12.

Newman, E.I. 1969b. Resistance to water flow in soil and plant. II. A review of experimental evidence on the rhizosphere resistance. *J. Appl. Ecol.* 16: 261–271.

Newman, E.I. 1976. Water movement through root systems. *Phil. Trans. R. Soc. B.* 273:463–478.

Nobel, P.S., and J. Sanderson. 1984. Rectifier-like activities of roots of two desert succulents. *J. Exp. Bot.* 35:727–737.

Oberlander, H.E. 1966. The relative importance of active and passive processes in ion translocation across root tissue. *In Limiting Steps in Ion Uptake by Plants from Soil.* International Atomic Energy Agency, Vienna, pp. 101–120.

Oosterhuis, D.M., and S.D. Wullschleger. 1987. Osmotic adjustment in cotton (*Gossypium hirsutum* L.) leaves and roots in response to water stress. *Plant Physiol.* 84:1154–1157.

Parker, W.C., and S.G. Pallardy. 1988. Leaf and root osmotic adjustment in drought-stressed *Quercus alba, Q. macrocarpa, and Q. stellata* seedlings. *Can. J. For. Res.* 18:1–5.

Parsons, L.R., and P.J. Kramer. 1974. Diurnal cycling in root resistance to water movement. *Plant Physiol.* 30:19–23.

Passioura, J.B. 1988. Water transport in and to roots. *Ann. Rev. Plant Physiol. Mol. Biol.* 39:245–265.

Philip, J.R. 1957. The physical principles of soil water movement during the irrigation cycle. *Third Congr., Int. Comm. Irrig. Drain.*, San Francisco. California Quest. 8, p. 8.125–8.154.

Puppi, G., and A. Bras. 1989. Nutrient and water relations of mycorrhizal white clover. *Agric. Ecosystems Environ.* 29:317–322.

Radin, J.W., and J.S. Boyer. 1982. Control of leaf expansion by nitrogen nutrition in sunflower plants. *Plant Physiol.* 69:771–775.

Reicosky, D.C., and J.T. Ritchie. 1976. Relative importance of soil resistance and root resistance in root water absorption. *Soil Sci. Soc. Am. J.* 40:293–297.

Richards, J.H., and M.M. Caldwell. 1987. Hydraulic lift: Substantial nocturnal water transport between soil layers by *Artemisia tridentata* roots. *Oecologia* (Berlin) 73:486–489.

Rosene, H.F. 1944. Control of water transport in local root regions of attached and isolated roots by means of the osmotic pressure of the external solution. *Am. J. Bot.* 28:402–410.

Saab, I.N., and R.E. Sharp. 1989. Non-hydraulic signals from maize roots in drying soil: Inhibition of leaf elongation but not stomatal conductance. *Planta* 179:466–474.

Shalhevet, J., E.V. Mass, G.J. Hoffman, and G. Ogata. 1976. Salinity and the hydraulic conductance of roots. *Plant Physiol.* 38:224–232.

Sharp, R.E., and W.J. Davies. 1979. Solute regulation and growth by roots and shoots of water-stressed maize plants. *Planta* 147:43–49.

Smith, R.C. 1960. Influence of upward water translocation on uptake of ions in corn plants. *Am. J. Bot.* 47:724.

Steudle, E. 1989. Water transport in roots. *In* B.C. Loughman, O. Gasparikova, and J. Kolek (Eds.). *Structural and Functional Aspects of Transport in Roots.* Kluwer Academic Publishers, Dordrecht, The Netherlands, pp. 139–145.

Steudle, E., and W.D. Jeschke. 1983. Water transport in barley roots. *Planta* 158:237–248.

Taura, T., Y. Iwaikawa, M. Furumoto, and K. Katou. 1988. A model for radial water transport across plant roots. *Protoplasma* 144:170–179.

Turner, N.C., and M.M. Jones. 1980. Turgor maintenance by osmotic adjustment: A review and evaluation. *In* N.C. Turner and P.J. Kramer (Eds.). *Adaptation of Plants to Water and High Temperature Stress.* John Wiley and Sons, Inc. New York, pp. 87–103.

Turner, N.C., E.-D. Schulze, and T. Gollan. 1985. The responses of stomata and leaf gas exchange to vapor pressure deficits and soil water content. II. In the mesophytic herbaceous species *Helianthus annuus*. Oecologia (Berlin) 65:348–355.

Turner, N.C., W.R. Stern, and P. Evans. 1987. Water relations and osmotic adjustment of leaves and roots of lupins in response to water deficits. *Crop Sci.* 27:977–983.

Tyree, M.T., and M.A. Dixon. 1983. Cavitation events in *Thuja occidentalis* L. Ultrasonic emissions from the sapwood can be measured. *Plant Physiol.* 72:1094–1099.

Tyree, M.T., and J.S. Sperry. 1989. Vulnerability of xylem to cavitation and embolism. *Ann. Rev. Plant Phys. Mol. Bio.* 40:19–38.

Tyree, M.T., E.L. Fiscus, S.D. Wullschleger, and M.A. Dixon. 1986. Detection of xylem cavitation in corn under field conditions. *Plant Physiol.* 82:597–599.

van Bavel, C.H.M., and J. Ahmed. 1976. Dynamic simulation of water depletion in the root zone. *Ecol. Modeling* 2:189–212.

van den Honert, T.H. 1948. Water transport in plants as a catenary process. *Disc. Faraday Soc.* 3:146.

Walker, R.F., D.C. West, S.B. McLaughlin, and C.C. Amundsen. 1989. Growth, xylem pressure potential, and nutrient absorption of loblolly pine on a reclaimed surface mine as affected by an induced *Pisolithus tinctorius* infection. *For. Sci.* 35:569–581.

Weatherly, P.E. 1963. The pathway of water movement across the root cortex and leaf mesophyll of transpiring plants. *In* A.J. Rutter and F.H. Whitehead (Eds.). *The Water Relations of Plants*. Blackwell, London.

Westgate, M.E., and J.S. Boyer. 1985. Osmotic adjustment and the inhibition of leaf, root, stem, and silk growth at low water potentials in maize. *Planta* 164:540–549.

Zhang, J., and W.J. Davies. 1987. Increased synthesis of ABA in partially dehydrated root tips and ABA transport from roots to leaves. *J. Exp. Bot.* 38:2015–2023.

Zhang, J., and W.J. Davies. 1989. Abscisic acid produced in dehydrating roots may enable the plant to measure the water status of the soil. *Plant Cell Environ.* 12:73–81.

Zhang, J., and W.J. Davies. 1990. Changes in the concentration of ABA in xylem sap as a function of changing soil water status can account for changes in leaf conductance and growth. *Plant Cell Environ.* 13:277–285.

Zhang, J., U. Schurr, and W.J. Davies. 1987. Control of stomatal behavior by abscisic acid which apparently originates in the roots. *J. Exp. Bot.* 38: 1174–1181.

Wheel-Induced Soil Physical Limitations to Root Growth

Ward B. Voorhees

I. Introduction 73
II. Soil Physical Limitations to Root Growth 74
III. Extent and Influence of Wheel Traffic 75
IV. Root Growth 80
V. Root Growth and Crop Yield 85
VI. Subsoil Compaction 88
VII. Conclusions 91
References 92

I. Introduction

Several aspects of the soil environment interact to affect root growth. Under field conditions, these factors are often very dynamic and seldom can it be determined, with a high degree of certainty, which factor was dominant. Eavis (1972) schematically diagrammed these interactions quite effectively. Mechanical impedance, or physical resistance to root elongation and function, is often cited as a limiting factor in crop growth and yield (Taylor, 1971). At the same time, it is recognized that the limiting factor is not physical resistance *per se* but is rather the restricted uptake of water and/or nutrients or inadequate gaseous exchange. Readers are referred to recent reviews such as Brown and Scott (1984), Hamblin (1985), and McCoy (1987) for detailed discussions on the soil physical effects on root growth.

Wheel traffic is one of the most common stimuli applied to agricultural fields and to which the response is often ignored or unknown. Practically every field operation, either on small experimental plots or large farm

Advances in Soil Science, Volume 19

fields, involves wheel traffic from a tractor or some other implement. If this wheel traffic did not produce changes in the physical soil environment, then it should not be of any scientific concern. However, wheel traffic can cause long-lasting changes in the physical soil environment (Voorhees, 1983) in ways that can significantly affect root growth and, therefore, potential grain yield.

This paper will restrict discussion to root growth in response to soil physical changes occurring as a result of mainly wheel-induced soil compaction and, to a lesser extent, soil tillage.

II. Soil Physical Limitations to Root Growth

Historically, physical limitations to root growth under field conditions in the United States have been observed most frequently in the warmer climatic regions. This was not likely due to temperature *per se* but was associated with low organic matter soils with non-expanding clays. Problems associated with genetically developed root-restrictive soil layers were compounded by cropping practices that often required numerous field operations and a high amount of wheel traffic, such as with cotton (*Gossypium hirsutum*) production (Taylor and Bruce, 1968). In sugar cane (*Saccharum officinarum* L.) production on lateritic soils, inordinate heavy wheel traffic significantly decreased root growth even though bulk densities were not extremely high, sometimes as low as $1.0\,kg\,m^{-3}$ (Trouse and Humbert, 1961). In this case, root growth was restricted because of pore rigidity and/or low air-porosity. Under cooler climatic conditions, where soils have a higher organic matter content and a more stable structure, root restrictions under field conditions are not as commonly observed and are not as easily related to yield responses.

Research reported by Miller et al. (1989), however, strongly suggested that soil, in general, is restrictive in allowing maximum crop yield. In their study, maize (*Zea mays* L.) was grown hydroponically out-of-doors adjacent to plants grown in natural soil conditions. Plants grown in soil achieved only 75 to 85% of the dry matter accumulated by plants grown hydroponically. There was no evidence that either soil temperature or nutrient uptake was a factor in these growth differences. Graham and Thurtell (1989) concluded that water stress was also not a factor. Miller et al. (1989) suggested that roots grown hydroponically lost less carbon than soil-grown roots. This is supported by research showing that a significant amount of carbon can be lost from the root into the rooting media (Smucker, 1982). Thus, it could be argued that most, if not all, soils present some degree of physical restriction to root growth and/or root function.

III. Extent and Influence of Wheel Traffic

Wheel traffic from various pre-planting operations such as fertilizer application, weed control, and secondary tillage is often visually obscured by the last tillage operation prior to planting. This "unseen" wheel traffic should no more be ignored in a plant growth study than should a previous "unseen" application of fertilizer in a nutrient uptake experiment.

The percentage of a farmer's field that is subjected to wheel traffic during the course of one growing season is significant. In a review by Hakansson et al. (1988), it was estimated that in modern farming with semi-random traffic patterns, the total area of a field covered by just the rear tractor wheels alone in one growing season is about twice the total field area. This can increase up to 3.5 to 4 times the total field area if harvest wheel traffic is included. Grimes et al. (1978) estimated that 70% of an alfalfa (*Medicago sativa* L.) field may be covered by wheel traffic during a single harvest. With maize planted in a ridge-till system, where all wheel traffic is confined to interrow areas, only 6 of every 18 interrows typically receives no wheel traffic, while the remaining 12 interrows have from 6 to 14 passes of a wheel when wheel traffic from all field operations is included (Parsons et al., 1984). Thus, wheel traffic can be a major influence affecting soil physical limitations to root growth.

A. Soil Properties

Several investigations have documented significant changes in soil environment as a result of wheel traffic that can affect root growth (Grimes et al., 1978; Trouse and Humbert, 1961; Voorhees, 1989). The two most commonly measured parameters are bulk density and penetrometer resistance. Data reported by Edwards et al. (1964) and Fehrenbacher et al. (1965) suggest a density of $1.8\,Mg\,m^{-3}$ in silt loam as the upper limit for maize root growth into discrete soil structural units. Voorhees et al. (1971) reported a similar upper limit for barley (*Hordeum vulgare* L.) root growth into clay loam aggregates.

Jones (1983) reviewed several data sets relating bulk density to root growth in soils with clay content from 5 to 65% and reported a negative relationship between the percentage of clay and critical bulk density (defined as the bulk density at which root growth is either maximum or at only 20 percent of maximum). Depending on the clay content, the upper critical bulk density ranged from 1.4 to $1.6\,Mg\,m^{-3}$. Even though this relationship was highly significant, abundant root growth was observed in some soils at a bulk density as high as $1.6\,Mg\,m^{-3}$, suggesting that roots were growing in fracture planes between peds or in biopores, neither of which are easily quantified by a bulk density-type measurement. Voorhees et al. (1978) showed that wheel traffic incurred during normal spring field operations on a well-structured clay loam soil in the northern

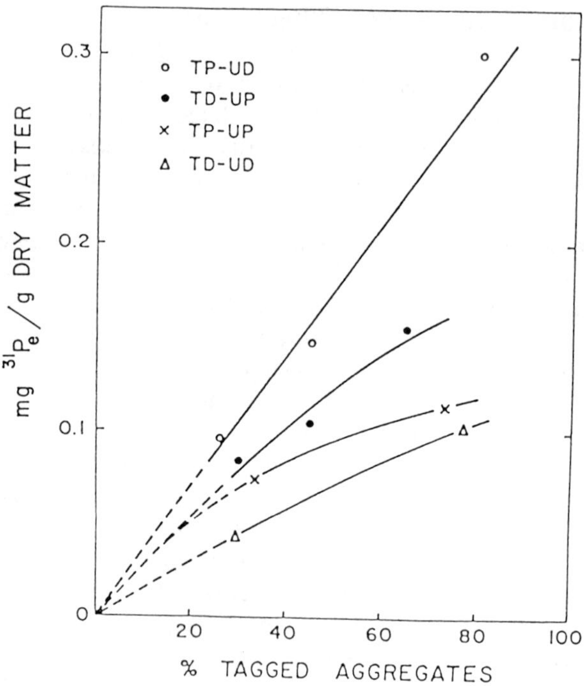

Figure 1. Uptake of tagged phosphorus from dense and porous soil aggregates. T = tagged; U = untagged; P = porous; D = dense. (from Voorhees et al., 1971)

Corn Belt can easily increase the bulk density in the surface 30 cm from 1.2 to 1.6 $Mg\,m^{-3}$. Bulk density values of 1.6 $Mg\,m^{-3}$ are fairly common at deeper depths in these soils, yet root growth may not always be seriously impeded. The root growth data reviewed by Jones (1983) were with soil water potentials near field capacity or wetter. Under rain-fed agriculture, many cultivated soils are seldom near field capacity, thus limiting the application of these critical bulk density values for defining root-restricting soil conditions.

Schuurman (1965) cautioned against a single-valued critical bulk density for root growth, citing the influence of surrounding soil conditions on mediating the growth of roots into a given soil volume. This was verified by Voorhees et al. (1971) who showed that barley root growth into and around individual soil aggregates, as indicated, in part, by the uptake of tagged phosphorus, depended on both the density of the aggregates and the heterogeneity of the aggregate density within the rooting media. Soil aggregates, 9 to 12 mm in diameter and of varying density, were formed from soil tagged with radioactive phosphorus. Roots were allowed to grow into containers filled with mixtures of different aggregate densities, tagged and untagged. Figure 1 shows the uptake of phosphorus from

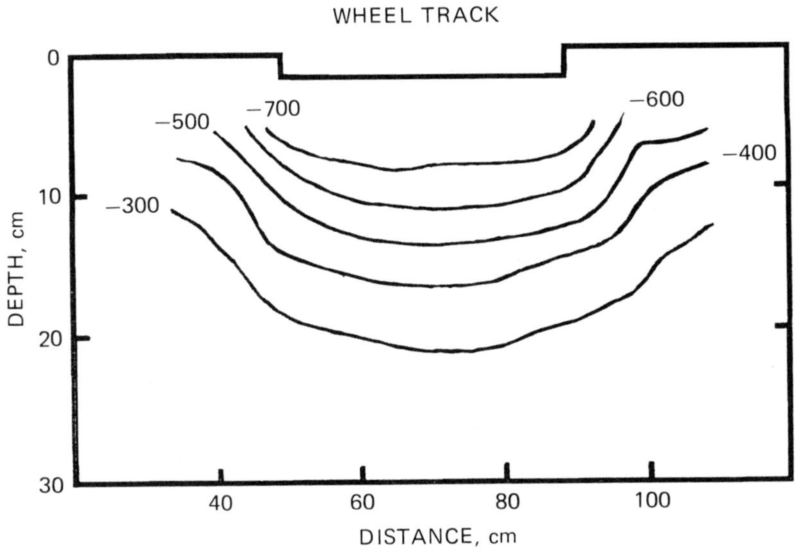

Figure 2. Two-dimensional plot of the hydraulic head after 14 days in a drying cycle. Wheel track was imposed under wet soil conditions. (from Reicosky et al., 1981)

tagged aggregates as a function of the percent of total aggregates tagged in various aggregate mixtures. The least amount of P uptake occurred when the rooting media consisted of all "dense" aggregates (1.88 $Mg\,m^{-3}$, curve TD-UD). When the rooting media consisted of all "porous" aggregates (1.52 $Mg\,m^{-3}$, curve TP-UP), P uptake increased. However, the greatest P uptake occurred when the rooting media consisted of a mixture of aggregate densities (curves TD-UP and TP-UD). The presence of some dense aggregates apparently forced more root growth into the porous aggregates, and the presence of some porous aggregates promoted more root growth around the dense aggregates.

B. Water Relationships

Increasing the bulk density of a soil also generally changes the soil water retention characteristics by decreasing the number of macropores and increasing the number of micropores. Since wheel traffic normally is not uniformly distributed over the entire field surface, non-uniform water movement and/or water use can be expected. This has been demonstrated under both laboratory and field conditions. Reicosky et al. (1981) measured soil water potentials in a two-dimensional pattern during a drying cycle from a box of soil that had a dense zone simulating an interrow wheel track. Isolines of equal soil water potential, as shown in Figure 2, suggest that the upward movement of water towards the

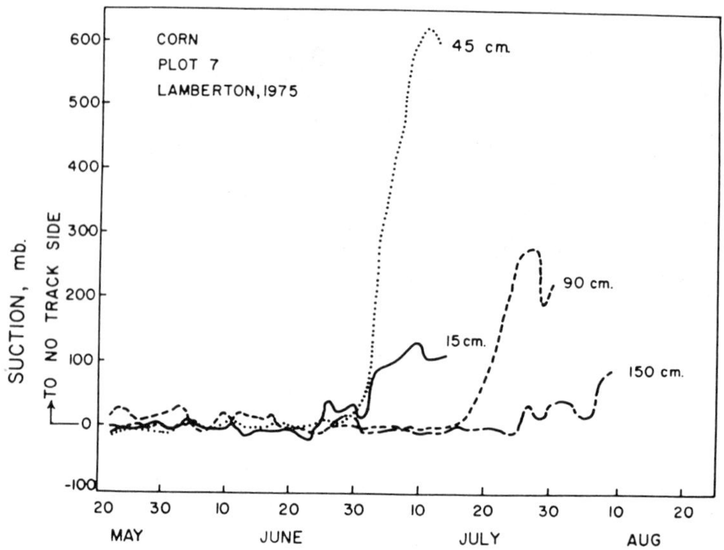

Figure 3. Lateral soil water potential gradients between tracked and nontracked side of maize row at specific depths during the growing season. (Voorhees, 1976)

evaporating soil surface should converge towards the compacted wheel-tracked soil. However, calculated rates of water movement through the compacted soil were much lower than through the non-compacted soil. In the absence of differential water extraction by root growth, a wheel-tracked interrow will thus likely remain wetter during a drying cycle than a non-tracked interrow.

The non-uniform lateral distribution of soil water potential in response to wheel traffic introduces another degree of heterogeneity in the field that can significantly impact root growth and function. Figures 3 and 4 show laterally non-uniform soil water potentials under field conditions in the presence of growing maize and soybean (*Glycine max* L. Merrill) roots, respectively. For both crops, soil water potentials were measured in the middle of the interrow on both sides of the plant; one interrow was wheel-tracked, the other was not. The plots were level and there was no obvious surface lateral movement of rainfall water between interrows. The values on the ordinate axes are the lateral differences between interrow soil water potentials measured at specific depths during the growing season. Positive values indicate a potential gradient for lateral water movement from the wheel-tracked side of the row to the non-tracked side.

There was no lateral water potential gradient at any depth for maize before June 30 (Figure 3). After June 30, there was a small gradient at the 15-cm depth toward the nontracked side of the row and a large

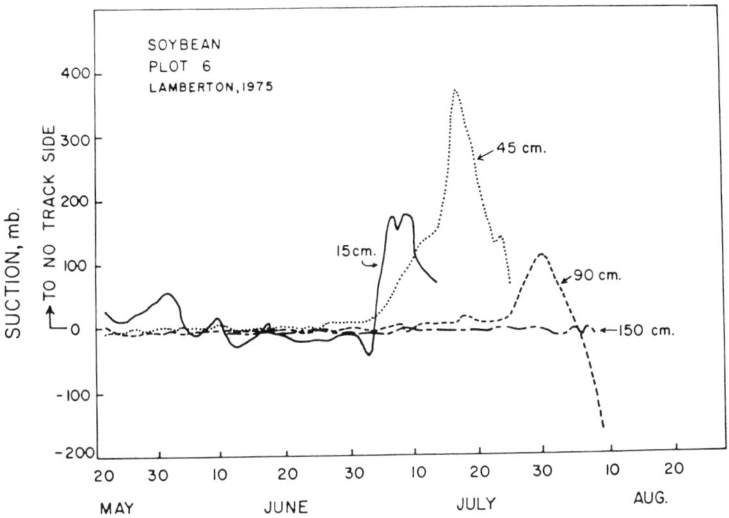

Figure 4. Lateral soil water potential gradients between tracked and nontracked side of soybean row at specific depths during the growing season. (Voorhees, 1976)

gradient in the same direction at the 45-cm depth. A lateral gradient towards the nontracked side of the row began at the 90-cm depth about July 15, followed 10 days later by a similar gradient at the 150-cm depth. This depth progression of a lateral gradient with time probably coincides with the advancement of the rooting front and suggests a faster depth-wise root growth rate on the nontracked side of the row and/or a more active root system in terms of water uptake. Soil bulk density at the 15-cm depth was 1.29 and 1.62 $Mg\,m^{-3}$ for the nontracked and tracked interrows, respectively. At deeper depths, there was essentially no difference in bulk density.

The sequence of lateral water potential gradients with depth and time for soybeans (Figure 4) was similar as that for maize, with the exception of two species-related trends. The time of maximum lateral difference at a given depth was delayed about 5 to 10 days in soybeans compared with maize, and soybeans did not apparently have any lateral difference in root growth and/or function at the 150-cm depth. This agrees with observations of shallower rooting for soybeans compared with maize in the northern Corn Belt (Allmaras et al., 1975). The interactions between wheel-induced soil properties, water potential, and root growth are very complex and not easily explained under field conditions. The lateral heterogeneity in root growth and water supply induced by interrow wheel traffic, especially in the surface 30 cm, can be a very important factor in fertilizer placement and management to optimize efficient chemical use.

Table 1. Bulk density at various depths as affected by tillage and wheel traffic (Voorhees et al., 1978).

	Moldboard plow		Reduced tillage	
Depth	no-track	wheel-track	no-track	wheel-track
cm		g/cm^3		
0–15	1.25	1.51*	1.29	1.54*
15–30	1.36	1.50*	1.43**	1.54*
30–45	1.48	1.50	1.53	1.55

* Within a tillage method, wheel-track is significantly higher than no-track at p = 0.01 as determined by student's t test.

** No-track reduced tillage is significantly higher than no-track moldboard plow at p = 0.01 as determined by student's t test.

IV. Root Growth

Tillage and wheel traffic often produce contrasting and complex effects on root growth, as illustrated in a data set reported by Voorhees (1989). Root length measurements were made in a field study conducted on a clay loam soil (Udic Haplustoll) in southwestern Minnesota over a three-year period. Maize and soybeans were grown with two primary fall tillage methods: moldboard plowing to a depth of about 25 cm and reduced tillage such as chisel or disking to a depth of about 20 cm. The controlled traffic concept was used whereby all wheel traffic was confined to the same interrows for the entire three-year period. A modified pinboard technique (Nelson and Allmaras, 1969) was used to extract soil monoliths with intact root systems from rows having no interrow wheel traffic and rows having interrow wheel traffic on both sides of the row. Root length density (kilometer of root per cubic meter of soil) was determined by a modified line-intersect method (Voorhees et al., 1980). A total of 27 and 33 monoliths of soybeans and maize, respectively, were taken over the three-year period. All monoliths were taken at about 60 to 70 days after planting.

Bulk density averaged over the three-year period is shown in Table 1 for the wheel-tracked and nontracked soil in the two tillage methods. Interrow wheel traffic significantly increased the bulk density compared to the nontracked soil in the surface 30-cm layer of both tillage methods. In addition, reduced tillage resulted in significantly higher bulk density than for moldboard plowing at the 15- to 30-cm depth when there was no wheel traffic. Similar results have also been reported in random traffic situations (Staricka et al., 1990). Table 2 shows penetrometer resistance averaged across tillage treatments for wheel-tracked and nontracked interrows. As with bulk density, interrow wheel traffic significantly increased penetrometer resistance in the 30 cm of soil. Both bulk density

Table 2. Maximum penetrometer resistance for various depth increments as affected by wheel traffic (Voorhees et al., 1978).

Depth	Maximum penetrometer resistance	
	no track	wheel-track
cm	kPa	
0–15	549	1409**
15–30	834	1154*
30–45	1229	1288
45–60	1255	1297

*, ** Wheel-track significantly higher than no-track at p = 0.05 and 0.01, respectively, as determined by student's t test.

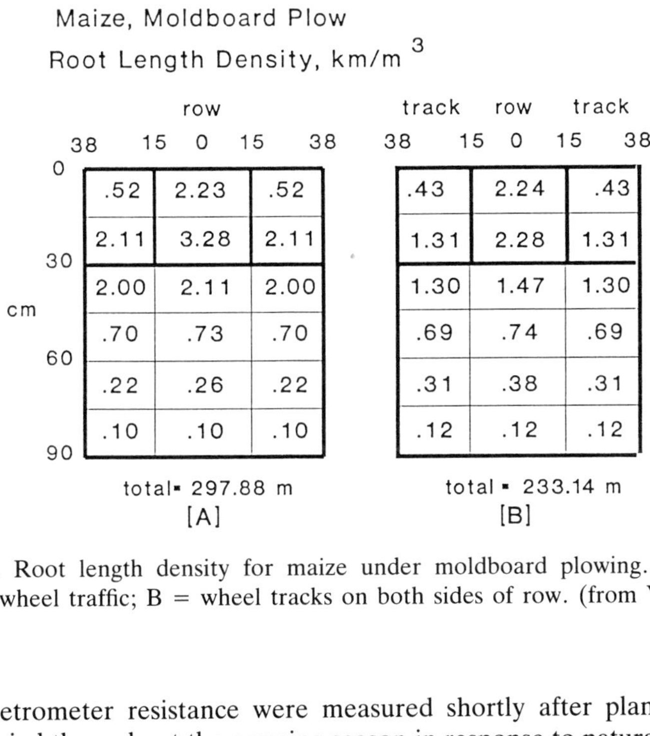

Figure 5. Root length density for maize under moldboard plowing. A = no interrow wheel traffic; B = wheel tracks on both sides of row. (from Voorhees, 1989)

and penetrometer resistance were measured shortly after planting and likely varied throughout the growing season in response to natural settling and soil moisture changes. Nevertheless, these data give an indication of the differences in porosity and strength characteristics due to wheel traffic and tillage that can affect root growth and function.

A. Maize Root Response to Wheel Traffic in Moldboard Plow System

Root length density at various depths and lateral distance from the row in the moldboard plow treatment is shown in Figure 5A and B for the

Maize, Reduced Tillage

Root Length Density, km/m ^{3}

[A]

cm	38–15	row (15–0–15)	15–38
0–15	.23	1.09	.23
15–30	1.15	1.69	1.15
30–45	1.16	1.12	1.16
45–60	.59	.52	.59
60–75	.26	.17	.26
75–90	.09	.09	.09

total = 171.46 m

[B]

cm	track (38–15)	row (15–0–15)	track (15–38)
0–15	.02	.77	.02
15–30	1.24	2.21	1.24
30–45	1.50	1.96	1.50
45–60	.75	.81	.75
60–75	.36	.38	.36
75–90	.11	.16	.11

total = 212.80 m

Figure 6. Root length density for maize under reduced tillage. A = no interrow wheel traffic; B = wheel tracks on both sides of row. (from Voorhees, 1989)

nontracked and wheel-tracked soil, respectively. Total root length for the nontracked profile was 297.88 m compared with 233.14 m for the profile with interrow wheel traffic. About 53% of the total root length was in the surface 30 cm of soil for both the tracked and nontracked profiles. There was little root growth in the surface 15 cm of the interrow regardless of traffic treatment. This layer was void of surface residue and was relatively dry. Thus, there was relatively little effect of interrow wheel traffic on root growth.

Interrow wheel traffic compaction increased root length below a depth of 45 cm even though there were no significant treatment-related changes in either bulk density or penetrometer resistance at these depths. Root length density at depths deeper than 45 cm are small and statistically insignificant, but may be extremely important for water uptake, especially under dry climatic conditions.

B. Maize Root Response to Wheel Traffic in Reduced Tillage System

Root data for the reduced tillage treatment are shown in Figure 6A and B for the nontracked and wheel-tracked profiles, respectively. There were 171.46 m of roots in the nontracked profile, 49% of which was in the surface 30 cm. In the wheel-tracked profile, there were 212.80 m of roots, but only 40% was in the surface 30 cm. Interrow wheel traffic drastically decreased root length in the surface 15 cm of the interrow but increased root length density throughout the remainder of the profile, especially under the row. Surface compaction increased total root growth by 24% in the reduced tillage treatment but decreased total root growth by 22% in the moldboard plow treatment.

These soils do not have a pronounced restrictive layer at the bottom of the tilled zone (20 to 30 cm). Moderate compaction within this tilled zone may decrease shallow root growth, but compensatory growth may occur at deeper depths. In another field study, a somewhat similar phenomenon was measured in response to subsoiling between maize rows to a depth of 51 cm. The interrow deep tillage (performed before planting) significantly reduced penetrometer resistance in the interrow but had no effect on interrow root length. However, root length in the row was increased by 88% even though there was no change in the penetrometer resistance within the row (Voorhees, 1989).

C. Maize Root Response to Tillage

There were also root growth responses that somewhat reflected the differences in tillage-induced soil physical properties. In the absence of wheel traffic, moldboard plowing resulted in increased root length density throughout the profile compared with reduced tillage (Figures 5A and 6A). This is in line with the lower bulk density and penetrometer resistance in the nontracked moldboard plowing (Tables 1 and 2). Hilfiker and Lowery (1988) also reported increased root length for moldboard plowing compared with reduced tillage in the absence of wheel traffic. Opposite results have also been reported. Sidiras and Kahnt (1988) compared root growth under moldboard plowing and rototilling. Rototilling resulted in more root growth in the subsoil even though the soil density in the tilled layer was higher. They attributed this enhanced deeper root growth to higher bio-pore area in the surface soil layer under rototilling. Similar results were reported by Dowdy et al. (1988), comparing root growth between moldboard plowing and reduced tillage. An important point is that bulk density measurements by themselves can not always explain how soil physical conditions affect root growth.

When there was interrow wheel traffic, differences in tillage-induced soil physical properties decreased, and moldboard plowing resulted in more root growth than reduced tillage only in the surface 30 cm (Figures 5B and 6B).

D. Soybean Root Response to Wheel Traffic in Moldboard Plow System

Vertical and horizontal root length distribution for soybeans under moldboard plow is shown in Figure 7A and B for the nontracked and wheel-tracked profiles, respectively. Total root length for the nontracked profile was 147.52 m, with 38% in the surface 30 cm. Interrow wheel traffic increased total root length to 187.24 m, with 49% in the surface 30 cm.

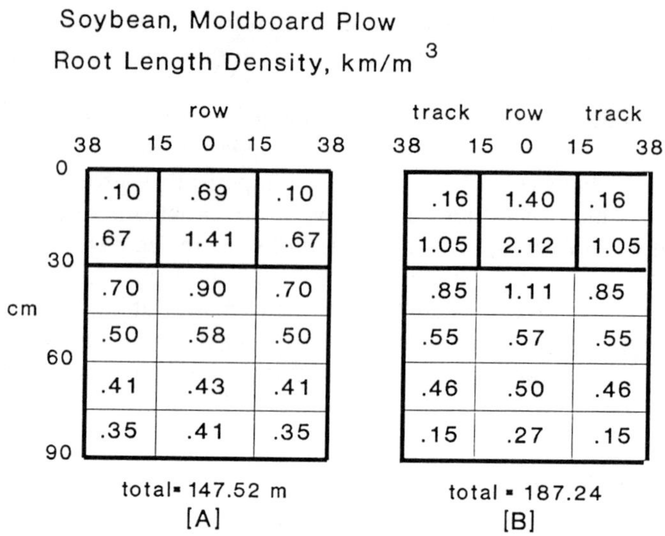

Figure 7. Root length density for soybeans under moldboard plowing. A = no interrow wheel traffic; B = wheel tracks on both sides of row. (from Voorhees, 1989)

The greatest relative increase in root length density in response to interrow wheel traffic was in the surface 30 cm under the row where the root length almost doubled. Root length density in the surface 30 cm of the interrow was also increased by interrow wheel traffic even though bulk density and penetrometer resistance also increased.

E. Soybean Root Response to Wheel Traffic in Reduced Tillage System

With reduced tillage, interrow wheel traffic decreased total root length from 147.88 m to 94.04 m (Figure 8A and B). This decrease was relatively uniform at all depths. Both traffic treatments had about 43% of the total root length in the surface 30 cm.

In the absence of interrow wheel traffic, total root length was not affected by tillage method, 147.52 m and 147.88 m for plowing and reduced tillage, respectively. The distribution, however, of roots was affected. Plowing resulted in increased root length density under the row and in the 30 to 45 cm depth, with the remainder of the profile generally decreasing (Figures 7A and 8A).

With interrow wheel traffic, plowing significantly increased root length density throughout the profile, especially in the 15- to 45-cm layer (Figures 7B and 8B).

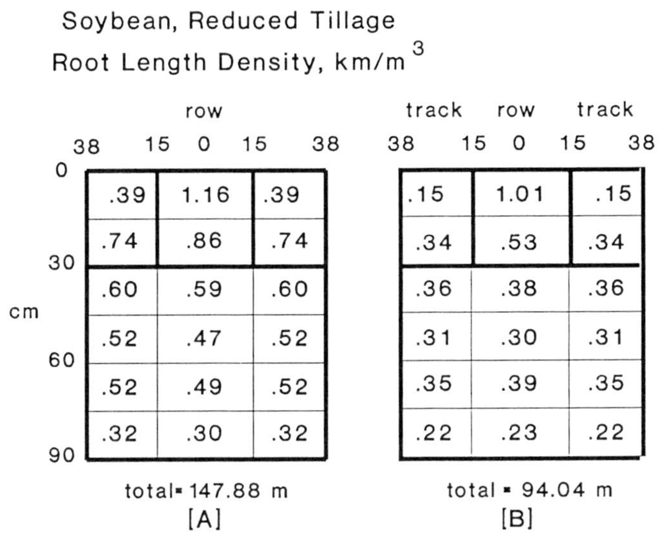

Figure 8. Root length density for soybeans under reduced tillage. A = no interrow wheel traffic; B = wheel tracks on both sides of row. (from Voorhees, 1989)

V. Root Growth and Crop Yield

As reviewed by Tardieu and Manichon (1987), water and nutrient uptake may be significantly influenced by the spatial arrangement of individual roots. While they were referring to a micro-scale (mm), the macro-scale (cm) variability discussed herein may also be important in water and nutrient uptake and, consequently, yield. The root data discussed in Figures 5 to 8 are point measurements in time and cannot be used to delineate cause-and-effect with respect to final yield. But since the root systems were sampled about 60 to 70 days after planting, the data may provide a first approximation (integrated over time) of the relationships between wheel-induced soil compaction, root growth, shoot growth, and yield.

Total shoot dry matter was measured at the same time the root data in Figures 5 to 8 were collected. Maize shoot dry matter was not obviously related to root length at that point in the growth stage. It was, however, directly related to root weight as shown in Figure 9. Approximately three grams of shoot biomass were generated for each gram of root biomass produced. Qualitatively similar results were obtained for soybeans (data not shown).

When maize and soybean data were pooled, shoot biomass was most closely related to root biomass in a volume of soil 30 cm deep and 15 cm

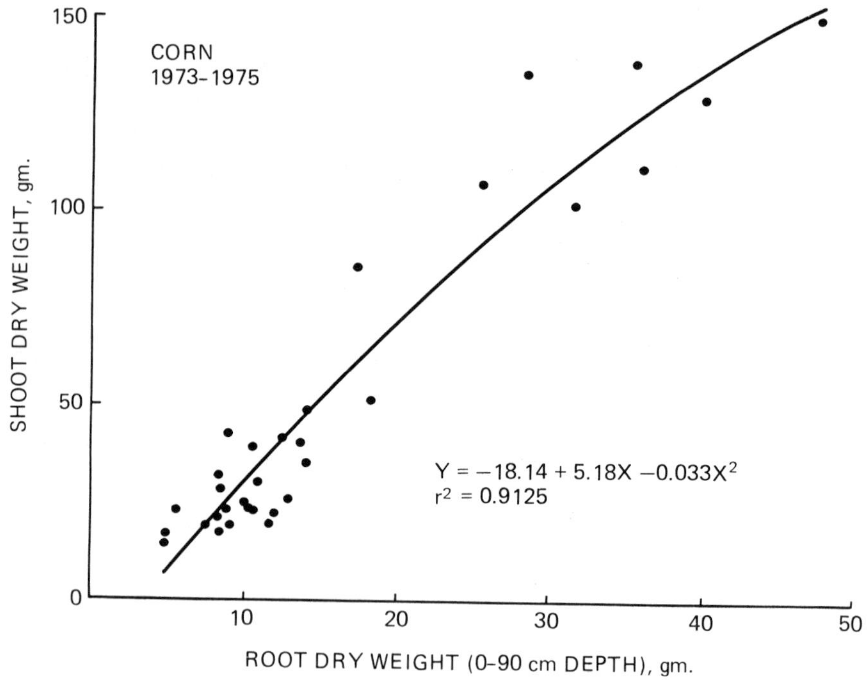

Figure 9. Maize shoot biomass as a function of total root biomass to 90-cm depth. Lamberton, Minnesota. (Voorhees, unpubl. data)

on either side of the row, as shown in Figure 10. The scatter of data suggests a near-linear relationship of about 8 g of shoot biomass produced per gram of root biomass up to about 10 g of roots, beyond which the shoot/root biomass ratio decreases to about three. In both Figures 9 and 10, maize root weight values extended over a wide range, whereas soybean data are clustered between 6 and 12 g of root dry matter in Figure 10. Interrow wheel traffic compaction had a greater effect on maize root biomass (and shoot biomass) than on soybean root biomass.

In monocotyledons "... the development of the mature root system depends largely on the sequential initiation of new nodal root axes ..." (Russell, 1977), which occurs primarily in the 30-cm cube of soil directly under the maize plant. Thus, the amount of root growth in this 30-cm cube of soil may be related to the final yield, especially for maize. This is shown in Figure 11 where the relative grain yield is plotted as a function of the portion of the total root length in the 30-cm cube of soil directly under the plant. Yield data associated with the four compaction-tillage treatments in Figures 5 to 8 were averaged over the three-year period. For a given crop, the maximum grain yield was given a relative value of one. It should be noted that the ordinate values are not necessarily

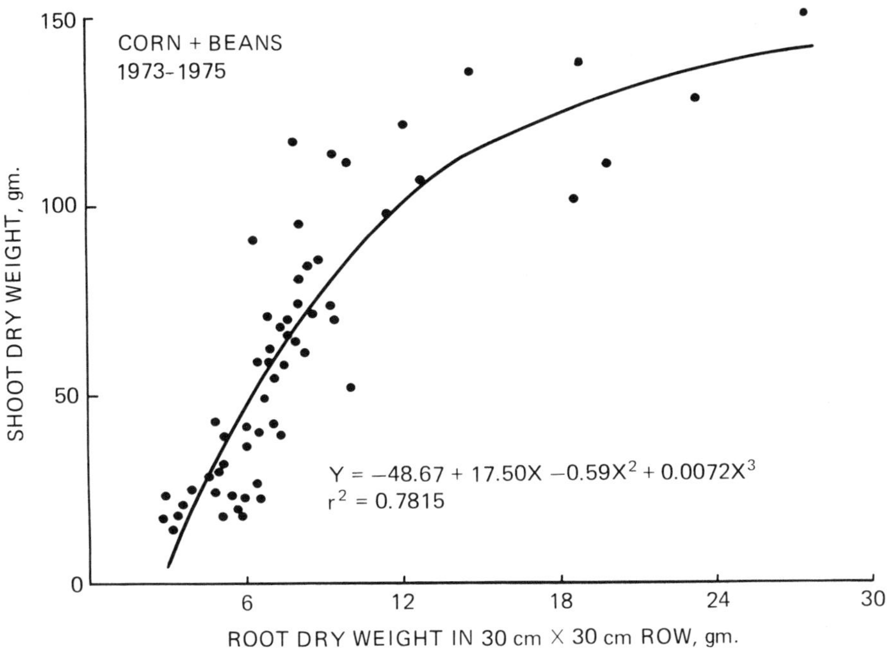

Figure 10. Shoot biomass of maize and soybeans as a function of root biomass in 30-cm cube of soil under the plant row. Lamberton, Minnesota. (Voorhees, unpubl. data)

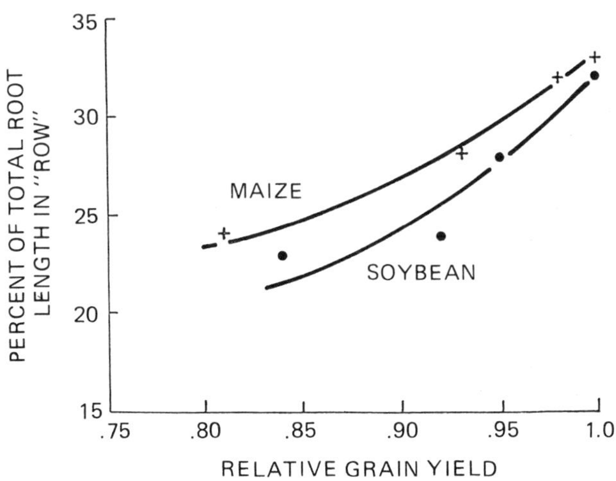

Figure 11. Grain yield of maize and soybeans as affected by root length in 30-cm cube of soil under the plant row. Lamberton, Minnesota. (Voorhees, unpubl. data)

increasing amounts of root growth, but are rather increasing percentages of a root system's total length being in the 30-cm cube of soil under the plant. These data suggest that a given plant's capacity to yield may be increased by concentrating the plant's total root system in the 30-cm cube of soil under the plant. Obviously, there is a limit to this reasoning, especially under rain-fed conditions. But, in principle, it agrees with more definitive field data reported by Allmaras and Nelson (1971) showing a trend for increasing maize grain yield as root growth in the shallow depths increased (this in response to temperature and soil water potential treatments). The growing season precipitation for the years in which the data in Figures 5 to 11 were obtained was below the long-term average. Since root growth patterns are very sensitive to climate-related soil conditions (Allmaras et al., 1975; Allmaras et al., 1987), caution must be exercised in extrapolating the above relationships beyond the original site-specific soil and climatic conditions.

VI. Subsoil Compaction

Subsoil compaction (defined as compaction deeper than the depth of normal tillage) is receiving increased attention because of the increasing size and weight of farm machinery (Voorhees et al., 1986). Readers are referred to discussions by Hakansson et al. (1987, 1988) for an international perspective of subsoil compaction. Crop yield response to subsoil compaction in North America has been reported by Gameda et al. (1987), Voorhees et al. (1989), Johnson et al. (1990), and Lowery and Schuler (1991). The collective data from all of these studies show that wheel traffic from machinery with axle loads in excess of 9 Mg (10 tons, typical for harvest and transport machinery) can cause increases in bulk density and penetrometer resistance deeper than 30 cm. Furthermore, these changes in soil physical properties persist in spite of annual freezing and thawing and, depending on soil type, climatic conditions, and crop species, can result in significant and long-lasting crop yield decreases.

None of the above studies included direct measurements of root growth. However, water use patterns can provide an indirect indication of rooting pattern and/or root activity. Again, results from a study in Minnesota will be used as a case example. In this field study, wheel traffic from an 18-Mg (20-ton) axle load (typical for a loaded eight-row combine) was applied to the entire plot surface of a relatively wet clay loam soil in the fall of 1981. This traffic caused significant changes in soil physical properties to a depth of 60 cm. The surface 25 cm was then intensively tilled to remove surface compaction. All wheel traffic in subsequent years was restricted to not exceed 4.5 Mg (5 tons) per axle. The control treatment never received traffic heavier than 4.5 Mg per axle. Maize and soybeans were grown on all plots for the following five years.

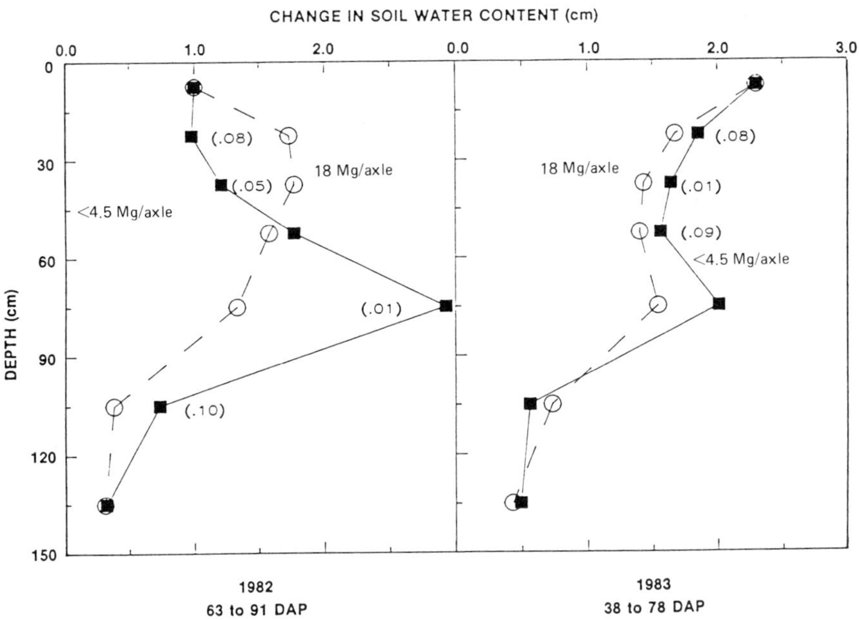

Figure 12. Change in soil water content under maize as affected by subsoil compaction. Squares are 4.5 Mg/axle load traffic (no subsoil compaction); circles are 18 Mg/axle traffic (compaction to 60 cm). Numbers in parenthesis indicate level of significance. Waseca, Minnesota. (after Voorhees et al., 1989)

Since 1987, only maize was grown. Detailed discussions are in Voorhees et al. (1986, 1989).

The 18 Mg/axle subsoil compaction treatment significantly decreased maize grain yield 30 and 12 percent in 1982 and 1983, respectively, the first two growing seasons after the initial application of high axle load wheel traffic. Significant differences in soil water loss were also measured during those two years and are shown in Figure 12. Over the period from 63 to 91 days after planting (DAP) in 1982, most of the soil water loss in the 4.5 Mg/axle treatment was from the 45 to 100 cm depth, whereas the 18 Mg/axle treatment lost the greatest amount of soil water from the surface 75 cm. The same trend existed in 1983 with the 4.5 Mg/axle treatment losing significantly more soil water in the surface 60 cm than the 18 Mg/axle treatment. The 18 Mg/axle treatment affected soil physical properties to a depth of 60 cm. These differences in water extraction reflect compaction-induced differences in water availability, rooting depth, and/or root activity.

More detailed field studies, such as those discussed by Logsdon et al. (1990), are required to define the relative role of the above factors. One of the complicating factors is that maximum rooting depth and/or root

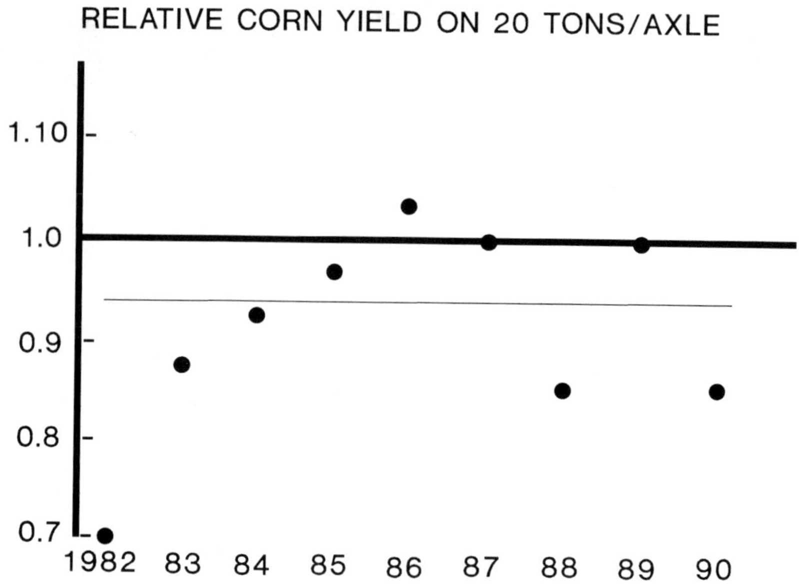

Figure 13. Maize yield from 18 Mg/axle traffic treatment relative to 4.5 Mg/axle treatment as a function of time. Waseca, Minnesota. (Voorhees, unpubl. data)

length density do not always directly relate to crop yield (as previously discussed with respect to shallow compaction). A 21-Mg axle load compaction treatment on a clay loam decreased the maximum rooting depth from 105 cm to 95 cm compared with a non-compacted treatment; however, maize grain yield was not affected (Voorhees, 1989). On sandy soils, maize, wheat (*Triticum aestivum*), and oat (*Avena sativa*) yields were directly related to increased rooting depth and root length density in the subsoil (Bennie and Botha, 1986; Jamieson et al., 1988). This complexity was illustrated by Noordwijk and Willigen (1987), who concluded that loosening a compacted subsoil would encourage rooting due to decreased mechanical resistance; however, decreased water storage and slower unsaturated hydraulic conductivity might discourage rooting depth.

One of the most serious consequences of subsoil compaction under current agriculture is that the effects on crop yield (and, presumably, on root growth, to a varying extent) can be long-lasting, even when subsoils are subjected to an annual freeze-thaw cycle. Maize yield data from the study associated with the data in Figure 12 is used to illustrate this. Figure 13 shows yearly maize yield from the 18 Mg/axle subsoil compaction treatment relative to the control treatment that never had wheel traffic heavier than 4.5 Mg/axle. It should be re-emphasized that the 18 Mg/axle treatment was applied only once, in the fall of 1981. The data for the first five years suggest an initial large yield depression (30%), followed by a gradual return to normal by 1986. However, during a stress year such as 1988 which was relatively dry or 1990 which was relatively wet, yield was significantly decreased by 15% each year. There is evidence that the yield decrease measured in the first year (1982) was partially due to incompletely ameliorated surface compaction and probably should not be considered as part of the long-term yield response. Averaging the remaining years gives an average long-term yield suppression of about 6%, due to a one-time application of 18 Mg/axle wheel traffic.

The potential seriousness of subsoil compaction is realized when farming trends are considered. The number of farms in the United States is decreasing but the average farm size is increasing. This, coupled with fewer farm operators, means that each operator must farm more land. The most obvious way to accomplish this is by using larger, and generally heavier, farm machinery. Conservation tillage practices are slowly decreasing the potential detrimental compaction effects from tractor wheel traffic by reducing the number of trips across the field. But the harvest and transport equipment continues to increase in size and weight, thus posing a continuing potential subsoil compaction problem.

VII. Conclusions

Wheel traffic often increases bulk density and penetrometer resistance in the surface 30 cm of soil. The extent to which this occurs depends on soil texture, soil water content, and pressure applied to the soil surface by the wheel. Increasing values of both bulk density and penetrometer resistance are generally considered as detrimental to root growth. This may be universally true if only an isolated individual root is considered. But when a root *system* is considered, especially under field conditions, factors such as heterogeneous soil conditions, compensatory root growth, and time depth-dependent changes in root function make it virtually impossible to generalize a relationship between soil conditions and root growth. Furthermore, parameters such as bulk density and penetrometer resistance do little to characterize soil water movement and the ability of roots to extract water, perhaps the most important factor in arable soils.

Keeping in mind the limitations of the above data to delineate cause and effect, the following trends with regards to wheel-induced soil compaction in the surface 30 cm of soil should be noted. Wheel traffic associated with tillage and planting in well-structured clay loam soils in the northern Corn Belt decreased maize root growth in the surface 30 cm but increased root growth in deeper layers. With moldboard plowing, surface compaction from wheel traffic generally decreased total root length while the opposite was true with reduced tillage. Soybean root growth response to surface compaction was generally opposite to that of maize root systems and probably reflects the difference between monocots and dicots in rooting form and function.

It is often assumed that compaction exerts its main influence on root growth by mechanically retarding root elongation. While this is definitely true in many soils, there are major land resource areas where this is probably not the main mechanism. The combination of water transmission to absorbing surfaces of the root system must be considered along with the ability of the root system to elongate to sites of water retention. It is important to consider root *systems*, realizing that a soil physical stimulus imposed on one portion of the root system can trigger a response in a different region of the total root system.

Long-lasting changes in soil physical properties in the subsoil due to high axle load wheel traffic present an additional set of problems. Depending on climatic conditions, yield responses to subsoil compaction may not be evident every year. From a practical farming perspective, an occasional compaction response may be very difficult to identify as such. Subsoiling to alleviate deep compaction and promote better root growth does not always produce the desired effect and may even be detrimental.

While increased root growth/depth would seem to be a viable goal as insurance for a water deficit situation, the above discussion points out the complexity of the relationships and precludes a general statement relating to the interactions between compaction-induced soil physical conditions, root growth, and crop yield. Much carefully designed research needs to be done under field conditions to delineate cause and effect.

References

Allmaras, R.R., and W.W. Nelson. 1971. Corn (*Zea mays* L.) root configuration as influenced by some row-interrow variants of tillage and straw mulch management. *Soil Sci. Soc. Am. Proc.* 35:974–980.

Allmaras, R.R., W.W. Nelson, and W.B. Voorhees. 1975. Soybean and corn rooting in southwestern Minnesota: II. Root distribution and related water inflow. *Soil Sci. Soc. Am. Proc.* 39:771–777.

Allmaras, R.R., S.C. Gupta, J.B. Sawn, and W.W. Nelson. 1987. Hydro-thermal effects on corn root development in a mollisol. *Trans. XIII Congress Intl. Soc. Soil Sci.* Hamburg, 1986, V:3–19.

Bennie, A.T.P., and F.J.P. Botha. 1986. Effect of deep tillage and controlled traffic on root growth, water-use efficiency, and yield of irrigated maize and wheat. *Soil Tillage Res.* 7:85–96.

Brown, D.A., and H.D. Scott. 1984. Dependency of crop growth and yield on root development and activity. *In* S.A. Barber and D.R. Bouldin (Ed.). *Roots, Nutrient and Water Influx, and Plant Growth.* Special Publication No. 49., Am. Soc. Agron., Madison, Wisconsin, pp. 101–136.

Dowdy, R.H., A.M. Bidwell, D.R. Linden, and R.R. Allmaras. 1988. Corn root distribution as a function of tillage and residue management. *Proc. of 11th Congress of Intl. Soil Tillage Res. Organ.*, Edinburgh, Scotland, 1:55–60.

Eavis, B.W. 1972. Soil physical conditions affecting seedling root growth. I. Mechanical resistance, aeration, and moisture availability as influenced by bulk density and moisture levels in a sandy loam soil. *Plant Soil* 36:613–622.

Edwards, W.M., J.B. Fehrenbacher, and J.P. Vavra. 1964. The effect of discrete ped density on corn root penetration in a planosol. *Soil Sci. Soc. Am. Proc.* 28:560–564.

Fehrenbacher, J.B., B.W. Ray, and W.M. Edwards. 1965. Rooting volume of corn and alfalfa in shale-influenced soils in northwestern Illinois. *Soil Sci. Soc. Am. Proc.* 29:591–594.

Gameda, S., G.S.V. Raghavan, E. McKyes, and R. Theriault. 1987. Subsoil compaction in a clay soil. II. Natural alleviation. *Soil Tillage Res.* 10:123–130.

Graham, M.E.D., and G.W. Thurtell. 1989. The effect of increased transpiration on photosynthesis of corn. II. Comparisons between hydroponically and soil-grown plants. *Agric. Forest Meterol.* 44:317–328.

Grimes, D.W., W.R. Sheesley, and P.L. Wiley. 1978. Alfalfa root development and shoot regrowth in compact soil of wheel traffic patterns. *Agron. J.* 70: 955–958.

Hakansson, I., W.B. Voorhees, and H. Riley. 1988. Vehicle and wheel factors influencing soil compaction and crop response in different traffic regimes. Soil Tillage Res. 11:239–282.

Hakansson, I., W.B. Voorhees, P. Elonen, G.S.V. Raghavan, B. Lowery, A.L.M. Van Wijk, K. Rasmussen, and H. Riley. 1987. Effect of high axle load traffic on subsoil compaction and crop yield in humid regions with annual freezing. *Soil Tillage Res.* 10:259–268.

Hamblin, A. 1985. The influence of soil structure on water movement, crop root growth, and water uptake. *Adv. Agron.* 38:95–158.

Hilfiker, R., and B. Lowery. 1988. Effect of conservation tillage systems on corn root growth. *Soil Tillage Res.* 12:269–284.

Jamieson, J.E., R.J. Morris, and C.E. Mullins. 1988. Effect of subsoiling on physical properties and crop growth on a sandy soil with a naturally compact subsoil. *Proc. 11th Congress of Intl. Soil Tillage Res. Organ.*, Edinburgh, Scotland, 2:499–503.

Johnson, J.F., W.B. Voorhees, W.W. Nelson, and G.W. Randall. 1990. Soybean growth and yield as affected by surface and subsoil compaction. *Agron. J.* 82:973–979.

Jones, C.A. 1983. Effect of soil texture on critical bulk densities for root growth. *Soil Sci. Soc. Am. J.* 47:1208–1211.

Logsdon, S.D., R.R. Allmaras, W.W. Nelson, and W.B. Voorhees. 1990. Persistence of subsoil compaction produced by heavy axle loads. *Agron. Abstr.* p. 320.

Lowery, B., and R.T. Schuler. 1991. Temporal effects of subsoil compaction on soil strength and plant growth. *Soil Sci. Soc. Am. J.* 55:216–223.

McCoy, E.L. 1987. Energy requirements for root penetration of compacted soil. *In* L.L. Boersma (Ed.). *Future Developments in Soil Science Research.* Soil Sci. Soc. Am., Madison, Wisconsin, pp. 367–377.

Miller, M.H., G.K. Walker, M. Tollenaar, and K.G. Alexander. 1989. Growth and yield of maize (*Zea mays* L.) grown outdoors hydroponically and in soil. *Can. J. Soil Sci.* 69:295–302.

Nelson, W.W., and R.R. Allmaras. 1969. An improved monolith method for excavating and describing roots. *Agron. J.* 61:751–754.

Noordwijk, M. van, and P. de Willigen. 1987. Agricultural concepts of roots: From morphogenetic to functional equilibrium between root and shoot growth. *Neth. J. Agric. Sci.* 35:487–496.

Parsons, S.D., D.R. Griffin, and D.H. Doster. 1984. Equipment wheel spacing availability and adaptations for ridge-planted corn and soybeans. *Am. Soc. Agric. Eng.*, Paper No. 84–1014, St. Joseph, Michigan.

Reicosky, D.C., W.B. Voorhees, and J.K. Radke. 1981. Unsaturated water flow through a simulated wheeltrack. *Soil Sci. Soc. Am. J.* 45:3–8.

Russell, R.S. 1977. *Plant Root Systems: Their Function and Interaction with the Soil.* McGraw-Hill, New York, 298 pp.

Schuurman, J.J. 1965. Influence of soil density on root development and growth of oats. *Plant Soil* 22:352–374.

Sidiras, N., and G. Kahnt. 1988. Biopores, rooting of maize, and physical soil properties as influenced by tillage systems. *Proc. 11th Congress of Intl. Soil Tillage Res. Organ.*, Edinburgh, Scotland, 1:115–121.

Smucker, A.J. 1982. Carbon utilization and losses by plant root systems *In* S.A. Barber and D.R. Bouldin (Ed.). *Roots, Nutrient and Water Influx, and Plant Growth.* Special Publication No. 49., Am. Soc. Agron., Madison, Wisconsin, pp. 27–46.

Staricka, J.A., P.M. Burford, R.R. Allmaras, and W.W. Nelson. 1990. Tracing the vertical distribution of simulated shattered seeds as related to tillage. *Agron. J.* 82:1131–1134.

Tardieu, F., and H. Manichon. 1987. Soil structure, rooting, and water uptake of maize. II. Growth and spatial distribution of the root systems. *Agronomie* 7:201–211.

Taylor, H.M. 1971. Effects of soil strength on seedling emergence, root growth, and crop yield. *In* K.K. Barnes, W.M. Carleton, H.M. Taylor, R.I. Throckmorton, and G.E. Vanden Berg (Eds.). *Compaction of Agricultural Soils.* Am. Soc. Agric. Eng. Monograph, St. Joseph, Michigan, pp. 292–305.

Taylor, H.M., and R.R. Bruce. 1968. Effect of soil strength on root growth and crop yield in the southern United States. *9th Intl. Cong. Soil Sci. Soc. Trans.* 1:803–811.

Trouse, A.C. Jr., and R.P. Humbert. 1961. Some effects of soil compaction on the development of sugar cane roots. *Soil Sci.* 91(3):208–217.

Voorhees, W.B. 1976. Plant response to wheel-traffic-induced soil compaction in the northern Corn Belt of the United States. *Proc. 7th Conf. Intl. Soil Tillage Res. Organ.*, Uppsala, Sweden, 44:1–44:6.

Voorhees, W.B. 1983. Relative effectiveness of tillage and natural forces in alleviating wheel-induced soil compaction. *Soil Sci. Soc. Am. J.* 47:129–133.

Voorhees, W.B. 1989. Root activity related to shallow and deep compaction. *In* W.E. Larson et al. (Eds.). *Mechanics and Related Processes in Structured Agricultural Soils.* Kluwer Academic Publishers, Dordrecht, The Netherlands. pp. 173–186.

Voorhees, W.B., V.A. Carlson, and E.A. Hallauer. 1980. Root length measurement with a computer-controlled digital scanning microdensitometer. *Agron. J.* 72:847–851.

Voorhees, W.B., W.W. Nelson, and G.W. Randall. 1986. Extent and persistence of subsoil compaction caused by heavy axle loads. *Soil Sci. Soc. Am. J.* 50:428–433.

Voorhees, W.B., C.G. Senst, and W.W. Nelson. 1978. Compaction and soil structure modification by wheel traffic in the northern Corn Belt. *Soil Sci. Soc. Am. Jour.* 42:344–349.

Voorhees, W.B., M. Amemiya, R.R. Allmaras, and W.E. Larson. 1971. Some effects of aggregate structure heterogeneity on root growth. *Soil Sci. Soc. Am. Proc.* 35:638–643.

Voorhees, W.B., J.F. Johnson, G.W. Randall, and W.W. Nelson. 1989. Corn growth and yield as affected by surface and subsoil compaction. *Agron. J.* 81:294–303.

Soil Chemical Factors Limiting Plant Root Growth

Charles D. Foy

I. Introduction 97
II. Soil Chemical Factors 98
III. Solving the Shallow Rooting Problem 116
IV. Challenges for Future Research 129
V. Summary 130
VI. Acknowledgments 131
References 131

I. Introduction

In most cultivated soils, the plow layer is treated periodically with fertilizers and lime to correct obvious problems of mineral element deficiency or toxicity. Hence, most plant rooting difficulties are imposed by the subplow layer of the soil profile. Such problems may be physical (plow pans or genetic hardpans) or chemical (mineral element toxicities or deficiencies) in nature or both. In any event, the problem is often manifested as drought and poor use of subsoil nutrients associated with shallow rooting.

Any mineral element may be present in deficient or toxic concentrations for a particular plant and, subsequently, limit the growth of roots and tops. Soil chemical factors that could limit plant growth include deficiencies or unavailabilities of the essential nutrient elements N, P, K, Ca, Mg, Mo, S, Cu, Fe, B, Zn, and Mn; toxicities of Al, Mn, or other metallic cations; salinity; and perhaps even a deficiency of Si, which is regarded as essential for some plants and beneficial to others.

1992 by Springer-Verlag New York Inc.
Advances in Soil Science, Volume 19

Perhaps the most important chemical limitation to root penetration and proliferation is that imposed by strongly acidic subsoils having pH values below 5.0 (Doss and Lund, 1975). Hence, this review will emphasize the effects of the acid soil complex on plant growth and means by which the problems may be alleviated. Soil chemical factors that limit root growth in acid subsoils can, of course, also limit crop production in surface soils that are infertile and inadequately limed. The components of the soil acidity complex have been thoroughly discussed in earlier publications (Foy, 1984; Kamprath and Foy, 1985). Growth-limiting factors associated with this complex include toxicities of H, Al, Mn, and other cations; deficiencies of Ca, Mg, P, Mo, and Si; and various interactions among these factors. The situation is further complicated by interactions of these factors with drought and with plant genotypes.

II. Soil Chemical Factors

A. Low pH (H ion Toxicity)

Direct effects of the H ion on plant growth are difficult to assess because, at soil pH values where it is potentially harmful, Al, Mn, and other mineral elements may be present in toxic concentrations, and the availabilities of essential elements, particularly Ca, Mg, P, Mo, and Si, may be suboptimal. In most acid soils (pH above 4.0), Al and Mn toxicities are probably more important than H ion toxicity in limiting the growth of plants, particularly the non-legumes. However, H ion toxicity may restrict the survival and activity of rhizobia and other soil microorganisms (Moore, 1974; Kamprath and Foy, 1985).

Because the effects of H ions are confounded with those of other factors in acid soils, investigators have resorted to nutrient solutions or sand cultures to study the effects of low pH. In general, solutions having pH values below 4.0 reduce root growth (Arnon et al., 1942; Islam et al., 1980). In contrast, root hair length was increased by 40% when the solution pH was increased from 5.5 to 7.2 (Ekdahl, 1957). Kerridge (1969) reported that wheat (*Triticum aestivum* L.) root length was slightly greater at pH 5.0 than at pH 4.0. Howard and Adams (1965) found that a solution pH below 4.2 reduced the growth rate of primary cotton (*Gossypium hirsutum* L.) roots.

Excess H ions compete with other cations for root absorption sites, interfere with ion transport, and cause root membranes to become leaky. Thus, roots can lose previously absorbed cations as well as organic substances (sugars, amino acids), and prolonged exposures to low pH may reduce their capacities for subsequent absorption of nutrients. For example, a solution having a pH below 4.0 markedly enhanced the loss of organic substances from cotton radicles, and this effect was reversed by raising the pH of the solution (Christiansen et al., 1970). Pre-treatment of barley roots with HCl at a pH below 5.0 caused a pronounced loss of K (Hussain et al., 1954). This loss was prevented by the presence of the polyvalent cations Ca^{2+}, Al^{3+}, La^{3+}, and Ce^{3+}. Acid pre-treatment at pH values below 6.0 also markedly reduced the capacities of roots for subsequent uptake of K. Barley (*Hordeum vulgare* L.) roots exposed to pH 3.0 lost the majority of their K, Ca, P, and soluble N contents, and their abilities for subsequent K absorption were reduced.

Because of their effects on nutrient uptake and retention by plant roots, excess H ions can increase plant requirements for Ca and perhaps other nutrients in the growth medium. For example, soybean taproots growing in the nutrient solution portion of a split medium required higher Ca levels at pH 4.5 than at 5.6 (Lund, 1970). The H ion, within the pH range of 3.0 to 5.0, markedly reduced Ca uptake by excised maize roots (Maas, 1969). Excess H ions can also decrease the uptake of Mg (Arnon et al., 1942; Moore et al., 1961; Islam et al., 1980; Blamey et al., 1982), Mn (Robson and Loneragan, 1970), Zn (Rashid et al., 1976), and Cu (Bowen, 1969). Clearly-identifiable H ion toxicity is highly unlikely in higher plants grown on agricultural soils but could play a role when the same plants are grown on strongly acidic mine spoils (some with pH values below 3.0) and in acid sulfate soils. Wilkinson and Duncan (1989) concluded that H ion toxicity (pH below 4.8) was a major factor influencing the growth of sorghum (*Sorghum bicolor* [L.] Moench) on acid soils of Georgia; however, more recent evidence (Shuman et al., 1990) indicated that Al toxicity was the dominant growth limiting factor for sorghum in four acid soils of this state.

The H ion is probably most important in the growth of legumes without N fertilizer (Andrew, 1978). It can affect rhizobial survival and multiplication in soils, root infection and nodule initiation, legume rhizobial efficiency, and nodulation of the host plant. The root infection process apparently requires a higher pH than rhizobial survival (Munns, 1978; Carvalho et al., 1981; Richardson et al., 1988a,b). Franco and Munns (1982) found that decreasing the pH of nutrient solutions from 5.5 to 5.0 decreased the number of nodules formed by beans (*Phaseolus vulgaris* L.), but variable pH between 4.5 and 5.5 did not affect nodule growth and nitrogenase activity. They emphasized the importance of low pH in the early nodulation of bean. Several earlier investigators attempted to determine critical soil pH values for rhizobial activity. Bryan (1923a,b) reported that alfalfa (*Medicago sativa* L.) bacteria were killed at a soil pH of 5.0, those of red clover (*Trifolium pratense* L.) at pH 4.5 to 4.7, and those of soybean (*Glycine max* [L.] Merr.) at pH 3.5 to 3.9. The critical pH for nodulation of red clover was lower than that for alfalfa. Nodules formed at any pH where plants grew, but their numbers were greatest at a pH near 7.0. Sewell and Gainey (1930) concluded that, within the range of pH 4.5 to 7.0, soil pH was a minor factor in alfalfa growth if all necessary elements were present. Soybean rhizobia remained viable at soil pH values of 4.8 to 8.3, but soybean roots showed depressed nodulation within the range of 3.8 to 4.6 and were healthy at pH 5.6 (Doolas, 1930). Spurway (1941) concluded that the optimal soil pH was 6.2 for alfalfa, 6.5 for red clover and sweet clover (*Melilotus* sp. [L.] All.), 6.0 for soybean, 5.0 for cowpea (*Vigna unguicalata* [L.] Walp), and 4.5 for lespedeza (*Lespedeza stipulacea* Maxim.). In much of this earlier work, the results may have been due to Al more than H ion toxicity. Also, the rankings of species according to acid soil tolerance could be changed by changing the host plant cultivar and the bacterial strain, and results would also depend upon interactions among host plant and bacterial genotypes and environmental constraints like drought.

B. Aluminum Toxicity

Aluminum toxicity is probably the most important growth-limiting factor for plants in most strongly acidic soils and mine spoils (Foy, 1974a, 1984, 1988; Kamprath and Foy, 1985). The problem is particularly severe below pH 5.0 but has been reported to occur as high as pH 5.5 in kaolinitic soils (Foy, 1974a). Much of the poor root development and drought seen on soils having strongly acidic (below 5.0) subsoil layers is probably due primarily to Al toxicity which limits both rooting depth and degree of root branching. Soileau and Engelstad (1969) and Soileau et al. (1969) obtained evidence that chemical factors (Al-related) were more important than physical factors in limiting cotton root growth in an acid (pH 4.4) fragipan soil. The critical soil pH at which Al becomes soluble or exchangeable in toxic concentrations depends upon many other factors, including the predominant clay minerals, organic matter levels, concentrations of other cations, anions and total salts, and, particularly, the plant species or cultivar (Foy, 1984; Kamprath and Foy, 1985). Such complexity makes it difficult to devise a soil Al test that will accurately predict toxicity under all conditions. In addition to its role as a toxic cation, Al can also be toxic as the aluminate anion in alkaline fly ash deposits or bauxite residues at pH levels above about 8.5 (Jones, 1961; Fuller and Richardson, 1986).

Aluminum toxicity is a complex disorder which may be manifested as a deficiency of P, Ca, Mg, or Fe or as drought stress (Alam and Adams, 1980; Foy, 1984, 1988; Kamprath and Foy, 1985). In some plants, the foliar symptoms resemble those of P deficiency (overall stunting; small, dark green leaves; late maturity; purpling of stems, leaves, and leaf veins; and yellowing and death of leaf tips). In others, Al toxicity appears as an induced Ca deficiency or reduced Ca transport within the plant (curling or rolling of young leaves, inhibited growth of lateral branches, or a collapse of growing points or petioles). Excess Al has even induced Fe deficiency symptoms in rice (*Oryza sativa* L.), sorghum, and wheat (IRRI, 1974; Clark et al., 1981; Furlani and Clark, 1981; Foy and Fleming, 1982).

Aluminum injured roots are stubby and brittle (like cucumber pickles treated with "alum"). Root tips and lateral roots become thickened and may turn brown. The root system as a whole becomes corraloid, with many stubby lateral roots but reduced fine branching. Such roots are inefficient in absorbing both nutrients and water. Cotton plants affected by severe Al toxicity sometimes wilt in nutrient solutions, particularly under high greenhouse temperatures (C.D. Foy, unpubl. data, 1967). Young seedlings are more susceptible than older plants (Thaworuwong and van Diest, 1975). Aluminum apparently does not interfere with seed germination, but does impair new root development and seedling establishment (Nosko et al., 1988).

Aluminum is reported to interfere with cell division in root tips and lateral roots; increase cell wall rigidity by cross-linking pectins; reduce DNA replication by increasing the rigidity of the double helix; fix P in less available forms in soils and on plant root surfaces; decrease root respiration; interfere with enzymes governing sugar phosphorylation; decrease deposition of cell wall polysaccharides; decrease production and transport of cytokinins; modify structure and function of plant membranes; reduce water uptake; and interfere with the uptake, transport, and metabolism of several essential nutrients. However, in low concentrations, Al is beneficial to some plants (Foy, 1974a, 1984). The current state of knowledge concerning Al toxicity and differential Al tolerance in plants has been covered in several recent reviews (Anderson, 1988; Foy, 1988; Roy et al., 1988; Taylor, 1988; Hai et al., 1989).

C. Tests for Aluminum Toxicity in Soils

Conventional tests for Al toxicity in soils (pH, exchangeable Al, percentage Al saturation of CEC (cation exchange capacity), and salt extractable Al) are not useful in predicting Al toxicity problems across a wide range of soils (Adams and Lund, 1966; Wright, 1989); the soil solution usually provides more accurate information (Baker, 1988; Bruce et al., 1988). However, for soils having similar parent materials and clay minerals, pH alone or absolute levels of Al extracted by KCl or other salts may be useful in predicting Al toxicity for a given plant (Blamey and Nathanson, 1977; McCormick and Amendale, 1983; Manrique, 1986). In general, a more useful predictor of Al toxicity is the percentage of the cation exchange capacity (CEC) occupied by Al (Evans and Kamprath, 1970; Blamey and Nathanson, 1977; Cregan, 1980; Farina and Channon, 1980; Kamprath and Foy, 1985). This Al saturation is determined by displacing soil Al with 1 *N* KCl or other neutral, unbuffered salts and expressing the Al as a percentage of the 1 *N* NH_4OAc CEC at pH 7.0 or as a percentage of the total exchangeable Ca, Mg, K, and Na (with NH_4OAc) plus KCl Al. This approach has many variations. To be most effective, the percentage Al saturation must be applied within a rather narrowly defined set of conditions because the critical Al saturation associated with toxicity varies with soil type and with plant species and genotype (Adams et al., 1967a,b; Foy, 1987).

Evans and Kamprath (1970) found that when the Al saturation reached 60% (KCl Al/KCl Al plus NH_4OAc exchangeable Ca, Mg, K, and Na) the Al concentration in the soil solution was generally greater than 1 ppm, and this relationship was fairly constant for kaolinitic, mixed, and montmorillonitic mineralogies. This situation was also associated with soil pH values of 5.0 or below. Based on these findings, Kamprath (E.J. Kamprath pers. comm., 1980) has suggested that Al saturations greater than 60% could be considered as growth-limiting for most crops and 10 to 20% for sensitive crops. Foy (1987) found that Al tolerant BH 1146 and Al sensitive Sonora 63 wheat cultivars differed significantly in growth over a range of 36 to 82% Al saturation in a Tatum (clayey, mixed, thermic typic Hapludult) subsoil. In this instance, Al saturation was KCl Al/KCl Al plus Ca plus Mg. The two cultivars could be separated under the following conditions: pH (1:1 soil-water) below 5.2; KCl Al levels greater than 2 cmol kg^{-1}; and Al saturations greater than 20%.

D. Identifying Aluminum Toxic Ionic Species in Acid Soils

Recently, there has been considerable effort to identify the toxic ionic species of Al at different pH levels in different acid soils. It is generally agreed that some measure of Al activity is more indicative of potential toxicity than concentrations of water soluble or exchangeable Al or percentage Al saturation of soil CEC (Adams and Lund, 1966). These investigators found that the critical Al activity which limited root growth in a soil solution was the same as in a nutrient solution ($>0.15 \times 10^{-5}\,M$). It is also recognized that the toxicity of even water soluble Al may be reduced by chelation of Al with organic acids or other ligands (Hue et al., 1986; Muchovej et al., 1988) and that this process is not taken into account when calculating theoretical activities of the Al^{3+} ion by GEOCHEM. Many investigators have reported that the calculated activity of the Al^{3+} ion is the single best indicator of toxicity, but some evidence exists for toxicity of the monomeric hydroxy cations $Al(OH)^{2+}$ and $Al(OH)_2^+$ and even for the polynuclear hydroxy complexes of Al. The $AlSO_4^+$ cation is generally considered non-toxic or much less toxic than the Al^{3+} ion. Recent evidence for these conclusions is summarized in paragraphs to follow.

Bell and Edwards (1986) concluded that only monomeric Al ions were toxic and that the sum of their activities gave the best index of toxicity. According to Wright et al. (1987), the calculated activities of Al^{3+}, using 8-hydroxyquinoline or aluminon Al as inputs, gave the best prediction of root and shoot growth of subclover (*Trifolium subterraneum* L.) and switchgrass (*Panicum virgatum* L.) on an acid soil that was low in dissolved carbon. In studies by Bruce et al. (1988), the activities of Al^{3+} and

$Al(OH)^{2+}$ gave the best association with the relative root length of soybean. Aluminum activities corresponding to 90% relative root yield were 4.0 and 0.5 μM for Al^{3+} and $Al(OH)^{2+}$, respectively. Their results suggested that $Al(OH)_3$, $Al(OH)_2{}^+$, and $AlSO_4{}^+$ were nontoxic. Another finding was that soil solution pH values measured in water were more sensitive indicators of root growth than those in 0.01 M $CaCl_2$. Kinraide and Parker (1987) found that root elongation by Al sensitive Tyler wheat was inhibited by Al^{3+} or mononuclear species in equilibrium with it but not by $AlSO_4{}^+$. They concluded that $AlSO_4{}^+$ was either nontoxic or tenfold less toxic than Al^{3+}. In a subsequent study, Parker et al. (1989) concluded that the calculated activity of Al^{3+} was the best single measure of Al stress, but they also found that increasing the solution pH without the formation of Al polyhydroxy complexes decreased Al toxicity; this suggested that other mononuclear forms had some toxic effects. In this connection, Moore (1974) found that the toxicity of soluble Al increased as solution pH increased from 4.0 to 4.5 and that the growth of wheat roots was well correlated with calculated concentrations of $Al(OH)^{2+}$. In addition, Al tolerant and sensitive wheat cultivars could be separated by root length vs. $Al(OH)^{2+}$ plots. In the studies of Parker et al. (1988), cited above, the sum of the calculated activities of mononuclear Al species was not as good an indicator of toxicity as that of Al^{3+}, and there was evidence that polynuclear hydroxy Al was also toxic.

In simple solutions containing only $CaCl_2$, $AlCl_3$, and HCl, Kinraide and Parker (1989) found no evidence that mononuclear hydroxy Al ions were toxic. Noble et al. (1988b) noted that Al toxicity was alleviated by $CaSO_4$ and attributed the effect, at least in part, to the formation of the less toxic $AlSO_4{}^+$ species. Wright et al. (1989) also demonstrated the nontoxicity of $AlSO_4{}^+$. Hai et al. (1989) concluded that free Al was mostly responsible for toxicity in rice but attributed part of the toxicity to $AlSO_4{}^+$.

Recent studies by Parker et al. (1989) illustrate the small amounts of Al required for toxicity. Aluminum activities as low as 3 or 4 μM (0.08 to 0.10 ppm) completely inhibited the growth of wheat roots in a $CaCl_2$ solution (15 μM Ca). An unexpected result in this study was that all plants were more tolerant to Al^{3+} than to Al_b (polymeric Al). Results suggested that the physiologies of the two toxicities were fundamentally different. Wheat cultivars were differentially tolerant to Al^{3+} but not to polymeric Al. Results of a chemical analysis indicated that the Al polymeric complexes were composed primarily of the so called Al_{13} polymer which may have formed during the experiment.

E. Aluminum Related To Drought and Root Growth

Aluminum toxicity in strongly acidic subsoils (pH 5 or below) is believed to be a major cause of shallow rooting and drought, particularly in the coastal plain of the southeastern United States. Excess Al can reduce water use by restricting both rooting depth and branching in acid subsoils (Kauffman and Gardner, 1978; Foy, 1988; Ritchey et al., 1988a; Goldman et al., 1989a,b). Rooting depth and branching pattern were not affected by irrigation alone. Doss and Lund (1975) found that reductions in the yields of seed cotton resulting from shallow rooting on acid subsoils were overcome by frequent irrigations of the surface soil. In addition, Al can damage the roots to the extent that they cannot absorb adequate water from moist soils or even nutrient solutions. Cotton plants were wilted when grown in nutrient solutions containing Al, particularly under high greenhouse temperatures (C.D. Foy, pers. obs.).

Aluminum sensitive cultivars of barley and sunflower (*Helianthus annuus* L.) were more tolerant to ordinary drought stress but less tolerant to Al-induced drought stress than Al tolerant cultivars of the same species (Krizek and Foy, 1988a,b,c; Krizek et al., 1988). In these studies, the effects of Al toxicity and drought stress were generally additive, and combining Al stress with drought stress magnified cultivar differences. When plants are grown in soils underlaid by acid (Al toxic subsoils), both ordinary drought tolerance and Al-related drought tolerance are frequently needed. Sorghum cultivars having acceptable levels of tolerance to both kinds of drought have been produced in Brazil (Borgonovi et al., 1984; Gourley, 1984). Acid soil tolerant (Al tolerant) sorghum genotypes have also been released in Georgia (Duncan et al., 1989), but their responses to Al × drought stress interactions have not yet been tested in the field.

Klimov (1985) studied Al × drought (osmotic) interactions in nutrient solutions containing PEG (polyethylene glycol) 400. The two factors were synergistic in reducing dry matter yields of oats (*Avena sativa* L.). Growth inhibitions greater than 50% were accompanied by gradual decreases in the water contents of plant tissues. Goldman et al. (1989a,b) reported Al × drought stress interactions in soybeans; the Al tolerant PI 416937 cultivar was more drought resistant than the Al sensitive Forrest. Kennedy et al. (1987) related Al tolerance in cotton cultivars to the ability to develop roots in acid subsoils and to extract moisture under field conditions. A drought resistant population of rye (*Secale cereale* L.), was more tolerant to Al stress than a drought susceptible strain. Individual plants selected from each population for Al tolerance were more tolerant in subsequent exposures to drought than were control plants (Bieka and Kuchler, 1980). Red oak (*Quercus rubra* L.) seedlings grown in complete nutrient solutions containing Al (0, 0.75, or 2 mol M^{-3}) and Ca (10 or 250 mol M^{-3}) showed reduced hydraulic conductivity and leaf area/root length ratio when the Al/Ca ratio was 2.9 or higher in the solutions. Hydraulic conductivity was positively correlated with leaf area/root length ratio.

F. Aluminum-Calcium Interactions and Root Growth

As mentioned earlier, a major constraint in diagnosing problems of restricted root growth is the separation of Al toxicity, Ca deficiency, and the interactions of the two factors. Much current research is emphasizing activity ratios of Ca/Al or Al/Ca as diagnostic tools. Recent literature on the subject is summarized in the following paragraphs.

Taproot elongation in Norway spruce (*Picea abies* [L.] Karst) and European beech (*Fagus sylvatica* L.) was reduced when the molar ratio of Al/Ca in solution was increased from one to two (Rost-Siebert, 1983). Root penetration of acidic soil profiles in Australia were in agreement with interacting effects of Ca (0.5 to 5 mM = 0.02 to 0.20 ppm) and Al (0 to 20 μM = 0.5 ppm) in separate culture solutions (Simpson et al., 1977). In solutions containing no Al, alfalfa showed no growth response to Ca (0.2 to 5 mM) but, with Al added at 12 to 20 μM (0.27 to 0.54 ppm), yield responses were obtained with increased Ca concentrations from 0.5 to 5 mM in solution. Lund (1970) found that root elongation in soybeans was reduced when the ratio of the activity of Al to Ca was greater than 0.02.

Field studies by Smit et al. (1987) indicated that the total root length of Douglas fir (*Pseudotsuga menziesii* [Mirbel] Franco) was affected by high exchangeable Al accompanied by low exchangeable Ca when the soil pH to a 5-cm depth was 3.8 or below. In damaged stands of Norway spruce, the molar ratio of Ca/Al in the soil solution was below 0.2 during most of the vegetative period. Needle concentrations of Al were not useful in estimating the levels of free Al in soils (Kayda and Zvacek, 1989). Aluminum injury has been associated with the displacement of Ca and Mg from the roots by Al (Gotbold et al., 1988) and with the decreased uptake of Ca, Mg, and P from deeper soil zones by beech and other trees (Bengtsson et al., 1988).

Hutton (1985) concluded that an unfavorable balance between Ca and Al saturation in Australian subsoils prevents deep rooting by perennials such as *Centrosema pubescens* Benth, and also of soybean and maize (*Zea mays* L.). Subsoils of the ultisols studied (50 to 100 cm) had pH values of about 4.5, 80 to 90% Al saturation of CEC (1.5 meq Al/100 g or more) and Ca saturations of 6 to 10% (2 meq/100 g of exchangeable Ca or less). Tolerances of Centrosema selections to acid oxisols were attributed to the capacity for continued absorption of Ca by roots, despite the high Al and low Ca saturations usually prevailing.

Wright et al. (1987) found that exchangeable Ca, soil Ca saturation percentage, and total soil solution Ca were positively correlated with snapbean (*Phaseolus vulgaris* L.) root and shoot growth in acid soils. Soil Al saturation percentage, total soil solution Al, and solution Al extracted with 8-hydroxyquinoline (15 sec.) were negatively correlated with growth. But the ratio of Ca/Al in the soil solution was more closely related to growth than the concentration of any single element in the soil solution.

Many horizons in this study of Appalachian soils had both high Al and low Ca levels and interactions between the two factors affected both yield and Ca uptake.

The molar ratio of Al/Ca was significantly higher in a soil solution from a site where red spruce (*Picea rubens* Sarg.) dieback occurred than on a site where trees were healthy (Shortle and Smith, 1988). Shoot and root growth of red oak grown in a nutrient solution were more closely related to 1 *N* Al^{3+}/Ca^{2+} than to any other solution variable; this ratio also most closely predicted leaf and root Mn and leaf Al and Al/Ca ratios (Kruger and Sucoff, 1989).

Bruce et al. (1988) found that suitable indices for predicting Ca limitations in soybean root growth were either the percentage Ca saturation of soil CEC or the Ca activity ratio in a soil solution. This ratio was defined as Ca activity/sum of Ca, Mg, Na, and K activities. Relative root length was 90% when the Ca activity ratio was 0.05 and the Ca saturation of CEC was 11%. Calcium activity and Ca concentration in solution were less useful and soil Al percentage saturation of CEC was not a good predictor of root growth.

Noble and Sumner (1988) obtained relatively poor correlation between soybean root growth and monomeric Al species and ascribed the results to Al × Ca interactions. However, considering both Al activities and Ca activities simultaneously produced an expression which was significantly correlated with root growth. The expression, termed Ca-Al balance (CAB), was derived as follows:

$$\begin{aligned} \text{CAB} = {} & 2\log(\text{Ca}^{2+}) - 3\log(\text{Al}^{3+}) + 2\log[\text{Al(OH)}^{2+}] \\ & + \log \text{Al(OH)}_2{}^{+}. \end{aligned} \tag{1}$$

Because this derived CAB expression was also highly correlated with concentrations of Ca in plant shoots, these investigators concluded that Ca uptake is controlled by Ca and monomeric Al species in solution. In another study (Noble et al., 1988a) which involved monomeric Al species, relative root length was most highly correlated with the calculated activity of Al^{3+} ($R^2 = 0.811$), followed by $Al(OH)^{2+}$ ($R^2 = 0.687$), and $AlSO_4{}^+$ ($R^2 = 0.338$) was least correlated with related root length. In this study, the derived CAB expression was the best predictor of root growth.

The Ca-Al balance (CAB) expression for predicting the effects of Al and Ca on root growth has been questioned by Kinraide and Parker (1989). In the CAB expression, given above, they argue that the activities of Al species are co-linear and that toxicity could be attributed to any Al species, rather than their combined effects. In attempts to evaluate the coefficient of CAB in either linear or sigmoidal form, Kinraide and Parker (1989) generally obtained non-significant values for the coefficients of log $Al(OH)^{2+}$ and log $Al(OH)_2{}^+$. They concluded that data from several experiments failed to support the model and that there was no justification for including all four terms in the equation. Their concern was that the CAB expression leaves the impression that the toxic species of Al have been identified and that their relative toxicities are known. In studies by Kinraide and Parker (1989) and Noble et al. (1988a) involving solutions containing constant total Al and Ca (only HCl added to control pH), the relative root elongation of four dicot species decreased with increasing pH values, suggesting that monomeric hydroxy Al ions were toxic. However, in studies by Parker et al. (1989) and Kinraide and Parker (1989), the relative root elongation rate in wheat decreased with increasing pH, which suggested that monomeric Al was not toxic. In contrast, Moore (1974) concluded that $Al(OH)^{2+}$ was toxic to wheat.

Fey et al. (1989) responded to the the criticisms of CAB as follows: the co-linearity of $Al(OH)^{2+}$ and $Al(OH)_2{}^+$ was considered; the Kinraide and Parker multiple regression analysis is invalid because variables are not independent as required; the alleviation of Al toxicity by other cations, such as H^+ and Na^+, may need to be considered; and valence weighting in the CAB expression suggested that $AlSO_4{}^+$ should be evaluated at activities three times those of Al^{3+} before its toxicity can be discounted (a condition not met in the Kinraide and Parker analysis). Fey et al. (1989) are currently evaluating the ionic potential charge/radius ratio, which may be a better weighting coefficient than valence alone. The CAB should be considered a prototypic expression; and more elaborate expression may be developed to predict root response to Al in more complex soil environments.

G. Manganese Toxicity

Manganese toxicity is probably the second most important growth-limiting factor (after Al toxicity) in many acid soils (Foy, 1973, 1984; Foy et al., 1988). Unlike Al, Mn seems to affect plant tops more directly than roots, but root damage follows when the toxicity is severe (Smith et al., 1987; Suresh et al., 1989). The solubility and potential toxicity of Mn to a given crop depends upon many soil properties, including total Mn content, pH, organic matter content, aeration, and microbial activity. Manganese toxicity generally occurs in soil with pH values of 5.5 or below, if the soil contains sufficient total Mn. However, it can also occur at higher soil pH values in poorly drained or compacted soils where reducing conditions favor the production of divalent Mn that plants absorb. Manganese toxicity has even occurred on calcareous soils at pH 8.1 where Fe-deficient flax (*Linum usitatissimum* L.) plants have created reducing conditions and/or low pH in their root zones (Moraghan, 1979). Thus, Mn can be toxic at a soil pH that is too high for Al to be soluble in toxic concentrations as a cation and too low for toxicity as the aluminate anion. Iron deficient snap beans grown in nutrient solutions absorb excess Mn and develop toxicity symptoms (Fleming, 1989).

Some soils simply do not contain sufficient total Mn to produce toxicity, even at a pH of 5.0 or below. Soils of the Atlantic coastal plain of the United States are lower in total Mn than those of the Gulf coastal plain (Adams and Pearson, 1967). Hence, at a given pH, Mn toxicity is less likely in the former than in the latter. However, even within the Gulf coastal plain, soils vary widely in Mn content (Sedberry et al., 1978). For a current view of knowledge concerning Mn in soils and plants, see the monograph edited by Graham et al. (1988) and earlier reviews by Foy (1973, 1984) and Foy et al. (1988).

Manganese toxicity occurred in alfalfa grown on several acid soils when the level of exchangeable Mn (1 *M* NH_4OAc, pH 7.0) was $50\,mg\,kg^{-1}$ or higher (Foy, 1964). Alfalfa yields were highest when these levels were reduced to less than $20\,mg\,kg^{-1}$ by liming. According to Sims and Wells (1985), soil acidity was the greatest single problem in the growth of burley tobacco (*Nicotiana tabacum* L.) in Kentucky. From one-half to two-thirds of tobacco soils were estimated to need lime each year. Manganese toxicity and Mo deficiency were considered the most important acid soil factors limiting the growth of tobacco in acid soils. Tobacco yield losses attributed to Mn toxicity alone were estimated at $30 to $40 million annually.

What is the evidence for Mn toxicity in subsoils where most rooting problems occur? Shuman et al. (1990) concluded that growth reductions in sorghum on four acid coastal plain soils of Georgia were due to Al toxicity and not Mn toxicity or Ca deficiency. Wright et al. (1987) found no evidence of Mn toxicity in snap beans grown on 55 horizons from 14 major, acidic Appalachian soils. Growth was most closely related to Al/Ca ratios in soil solutions. In subsequent studies, Wright et al. (1989) concluded that Mn toxicity was not the major limiting factor in the growth of subclover and switchgrass on 11 acid soil horizons in West Virginia. Measured and calculated values of soil solution Mn generally provided the best correlations with Mn uptake; root Mn concentrations were highly correlated with solution Mn. Manganese extracted with 0.01 *M* $CaCl_2$ was also significantly correlated with plant Mn concentrations and uptake; however, Al toxicity and Ca deficiency were judged more important in limiting growth. Singleton et al. (1987) concluded that toxic levels of Mn are likely in many topsoils and soil horizons with Mn concretions. Hue (1988) reported Mn toxicity in a manganiferous oxisol treated with a low Mn sewage sludge in Hawaii. Toxicity was thought to result from chelation of native Mn by organic matter in the sludge.

In nutrient solutions, the growth of lettuce (*Latuca sativa* L.) root and hypocotyl was reduced by Mn concentrations of 2, 4, and $6\,mM\,Dm^{-3}$ at pH values of 3.5, 4.5, and 5.5, respectively (Kummerova and Buresova, 1989). Mascarenhas et al. (1985) reported that as little as 2 ppm Mn in solution was toxic to Forrest soybean and that Mn tolerance rankings of cultivars in nutrient solutions agreed well with those on acid, Mn toxic soil. These investigators noted that Mn can be toxic at a higher pH than Al and, hence, more lime may be needed to prevent toxicity in Mn sensitive crops like soybeans. Because it is not generally economically feasible to lime the entire soil zones in which roots are desired, they advocated the development of Mn tolerant cultivars.

Jarrell (1989) found that sugar beets (*Beta vulgaris* L.) grown with saline irrigation water contained more than 1000 ppm Mn in their tops; he considered this situation potentially Mn toxic.

H. Calcium Deficiency

Absolute Ca deficiency, like H ion toxicity, is difficult to identify within the acid soil complex (Foy, 1974a,b, 1984, 1988; Kamprath and Foy, 1985). Most acid soils contain adequate total Ca for most plants (Kirkby and Pilbeam, 1984; Rorison and Robinson, 1984), and Ca deficiency symptoms are rarely seen in the field. Only in the most highly leached, acid, low CEC soils (sands, oxisols) would absolute deficiencies be likely. Even in cases where the deficiency symptoms are observed, they are more likely due to Al-Ca antagonism than to low Ca supply per se (Long and Foy, 1970). Levels of Ca required for essential growth functions are so low as to approach those of micronutrients (Wallace et al., 1966). Hence, the majority of the Ca in soils and in plants serves as an excluder or detoxifier of other elements, such as Al and heavy metals that might otherwise be toxic. Calcium has been reported to provide some protection against drought (seed treatment with $0.5\,M$ $CaCl_2$; Asghar et al., 1986), salinity (Katsuhara and Tazawa, 1987; Lynch et al., 1987; Maas and Grieve, 1987), mechanical stress (Jones and Mitchell, 1989), and Ni, Al, and other potentially toxic elements (Foy, 1984; Korner et al., 1987). Calcium is specifically required for peanut (*Arachis hypogaea* L.) fruiting, although peanut is classified as Al tolerant and can tolerate acid soils if Ca is supplied (Alva et al., 1989). For details concerning postulated critical Ca levels in soils and solutions and Ca requirements of plants, see earlier reviews by Foy (1974b), Hanson (1984), and Clark (1984).

Several investigators have recently reported evidence for absolute Ca deficiencies in some acid soils. Roots of many crop cultivars grew well in acid soils of pH 4.3 to 4.6 treated with $CaSO_4$ (Inoue et al., 1988). Blair et al. (1988) found that, in an acid oxisol, Leucaena (*Leucaena leucocephala* [*Lam*] deWit) did not respond to pH increases produced by $SrCO_3$ but did respond to $CaCO_3$. In an ultisol, growth was improved by increasing pH but not by the addition of $Ca(NO_3)_2$. The conclusion was that the oxisol was Ca deficient and the ultisol was Al toxic. However, the lack of response to $SrCO_3$ does not necessarily mean that the level of Ca in the original soil was limiting growth. The $SrCO_3$ could have produced a Sr-Ca antagonism which reduced the effectiveness of the native soil Ca. In our studies (Foy, 1974b), raising the pH of an ultisol with $MgCO_3$

induced Ca deficiency in barley. Wallace (1989) concluded that Ca deficiency limited the growth of snap beans in an acid clay loam at pH 3.3, but this pH is low enough that Ca deficiency symptoms could have been caused by H-Ca or Al-Ca antagonism rather than absolute low levels of Ca in the soil.

Bell et al. (1989) determined the external Ca requirements for the growth and nodulation of six tropical food legumes in flowing nutrient solutions. For example, the solution Ca concentration associated with maximum growth of cowpea was $12\,\mu M$. However, such results may have little relevance in acid soil solutions where Al is generally present to antagonize Ca.

Moore and Patrick (1989) concluded that Ca and Mg deficiencies may occur in rice in acid sulfate soils of Thailand. Again, it appears that any Ca and Mg deficiency problems encountered were induced by Al, or perhaps by H ions, if pH values were below 4.0. In a study of 4 topsoils and 17 subsoils of Australia, Bruce et al. (1988) concluded that all of the subsoils were deficient in Ca. Thus, the first limiting factor for root growth was thought to be Ca deficiency and not Al toxicity, in spite of a low pH and a high degree of Al saturation. In this instance, it is impossible to separate the Al and Ca effects; any Ca deficiency observed was almost certainly induced by Al. Njohn et al. (1987) claimed that Ca deficiency was an important growth limiting factor for maize on acid ultisols in eastern Nigeria. This conclusion was based upon the following observations: $CaSO_4$ depressed soil pH but significantly increased dry matter yields of both roots and tops; $MgCO_3$ eliminated exchangeable Al but maize showed severe symptoms of Ca deficiency; and $Ca(OH)_2$, which also raised the pH and eliminated exchangeable Al in addition to supplying Ca, increased yields. Results suggested that Ca deficiency was a more serious growth limiting factor than Al toxicity. If these results are valid, limestone, which is expensive in Nigeria, can at least be partially replaced by less expensive Ca-containing wastes such as cement furnace slag.

The results of Ritchey et al. (1982) support the hypothesis that root penetration in certain Brazilian oxisols (clayey B21 horizons) was limited by lack of Ca rather than Al toxicity. This conclusion was based upon the following evidence: such horizons had low levels of exchangeable Ca (0.2 meq/100 g or less) and no exchangeable Al was detected; in untreated soil, the growth of wheat roots stopped within a few days after seed germination; the addition of 0.10 to 0.15 meq Ca/100 g soil as calcium chloride, phosphate, or carbonate normalized seedling growth; calcium chloride decreased soil pH but still increased root growth; magnesium carbonate was not beneficial; and wheat leaves on untreated soil contained 0.05% Ca compared with 0.13% in those grown with supplemental Ca. These investigators identified several profiles in which the B21 horizon (beginning at the 40 to 50 cm depth) was free of exchangeable Al and contained only 0.1 to 0.2 meq Ca/100 g of soil. They reasoned that rooting depth could be increased by the application of soluble Ca to the plow layer and allowing the naturally high rainfall of the region to leach Ca into the desired rooting zone. In subsequent work, Ritchey et al. (1988a,b) have proposed seedling growth tests for Al toxicity and Ca deficiency in acid subsoils. To test for Al toxicity in a given crop, root growth in a soil treated with 0.1 meq Ca/100 g of soil is compared with that in a soil receiving sufficient lime to neutralize exchangeable Al. The response to lime indicates that Al is involved. Aluminum toxicity is also detected by comparing the growth of Al tolerant and sensitive species or cultivars on soil receiving 0.1 meq Ca/100 g of soil to correct absolute Ca deficiency.

The conclusions in the foregoing paragraph may be questioned as follows: low absolute Ca levels do not necessarily mean that low Ca limits growth, and low absolute levels of exchangeable Al do not necessarily mean that Al toxicity does not limit growth. Oxisols do have low CEC values, but as much as 90% of this may be exchangeable Al. It is difficult to imagine a soil that is sufficiently acidic to have a low Ca level but *no* exchangeable Al. Perhaps the method used to detect Al is not sufficiently sensitive to measure that Al. It takes very little Al to be harmful when Ca is low. Another problem is that the addition of even small amounts of Ca as $CaCl_2$ will reduce any Al-Ca antagonism present. Is the observed Ca deficiency in such soils Al induced or absolute?

Additional evidence of Ca deficient acid soils has been obtained in Alabama. Adams and Moore (1983) studied factors limiting root growth of cotton in soil horizons representing profiles of six major soil series of the coastal plain. Soils were sampled to a depth of one meter. The treatments were: *check*; $CaSO_4$ to correct Ca deficiency; MgO to correct Al toxicity and aggravate Ca deficiency; and $Ca(OH)_2$ to correct both Ca deficiency and Al toxicity. Based on these criteria, Ca deficiency was more widespread than expected. It was detected in all three Dothan (fine-loamy, siliceous, thermic plinthic Paleudult) horizons (A2, B1, Bt), in all three Poarch (coarse-loamy, siliceous, thermic plinthic Paleudult) horizons (B1, B21t, B22t), in Lucy (loamy, siliceous, thermic plinthic Paleudult) (A3), and in Malbis (fine-loamy, siliceous, thermic plinthic Paleudult) (B23t). Calcium deficiency was reported in both eluvial and illuvial horizons. Calcium deficiency occurred when the calculated Ca activity (GEOCHEM) was 0.27 m*M* or less and the Ca saturation of CEC was 17% or less. Calcium deficiency was indicated by straight, small diameter roots with brown tips, in contrast to the swollen, stubby, gnarled root pattern produced by Al toxicity.

I. Magnesium Deficiency

Magnesium deficiency may limit plant growth on acid, sandy, highly leached soils having low CEC values (and perhaps high Al saturations), acid soils low in available Mg and limed to near neutrality with calcitic lime, and soils treated with high levels of K (Kamprath and Foy, 1985). Critical Mg saturation levels have been reported as 5 to 10% of CEC (Adams and Henderson, 1962; McLean and Carbonell, 1972; Fox and Piekielek, 1984). According to Mayland and Grunes (1979), the Mg/K ratio should be at least 0.5. Like the uptake of Ca, that of Mg is antagonized by Al. Kurashima and Kimura (1987) found that herbage yield of cocksfoot (*Dactylis glomerata* L. Pers.) was closely related to exchangeable Al and decreased with decreasing pH below 5.0; Al interfered markedly with Mg uptake. Grimme (1984) concluded that the degree of Al toxicity in soybean was a function of the severity of Mg deficiency. For additional information, see earlier reviews by Clark (1982) and Foy (1984, 1988).

J. Manganese Deficiency

In the sandy coastal plain soils of the southeastern United States, liming to a pH near neutrality often produces Mn deficiency in some crops. These include soybeans in Georgia (Boswell et al., 1981), peanuts in Virginia (Hallock, 1979), and Bermuda grass (*Cynodon dactylon* [L.] Pers.) in Florida (Snyder et al., 1979). Soybeans develop Mn deficiency at pH values of 6.2 or above in the coastal plain of South Carolina and adjacent states. Manganese deficiency also occurs in soybeans in Ohio (Kroetz et al., 1977) and in wheat in Tanzania (Hoyt and Myovella, 1979).

K. Molybdenum Deficiency

Unlike other micronutrients, Mo is less available in acid soils than in those limed to near neutrality. Hence, in many soils, liming corrects Mo deficiency by increasing the solubility and plant availability of native Mo. However, some soils are absolutely deficient in Mo and liming does not solve the problem. For example, on certain Australian soils, soil acidity is associated with both Mn toxicity and Mo deficiency, and subclover responds to Mo even when the soils are limed (Coventry et al., 1987). Ferreiro et al. (1985) reported that Mo is absorbed on goethite, hematite, bayerite, and alpha Al_2O_3, and that this adsorption is at a maximum at a soil pH of 4.0.

In the United States, Mo deficiency occurs most frequently in the eastern states; crops affected include alfalfa, cauliflower (*Brassica oleracea* [Botrytis Group]), broccoli (*Brassica oleracea* [Italica Group]), soybeans, clover (*Trifolium* sp.) truck crops, and citrus sp. (Clark, 1982). Yellow leaf spot, a disorder in cashew (*Anacardium occidentale* L.) attributed to Mo deficiency, is corrected by foliar sprays of Mo (Subbiah et al., 1986). Molybdenum deficiency is apparently widespread in northern Alabama (Burmester et al., 1988). These investigators found that the critical soil pH for soybean varied widely on 15 field sites. Of these, 13 sites showed lime response; 12 sites showed a Mo response; five sites showed a response to Mo alone (indicating absolute Mo deficiency); and in six sites, liming corrected both Mo deficiency and Al toxicity. In these studies, the soil parameter best correlated with Mo deficiency was the Fe/Mo ratio in an ammonium oxalate extract; the critical value was 1540. Molybdenum is a component of at least five enzymes, including nitrogenase which is involved in microbial N_2 fixation (Clark, 1982, 1984; Marschner, 1986). The Mo requirement of root nodules in legumes and in non-legumes (alder) is relatively high; hence, plants that depend upon N_2 fixation are particularly stimulated by Mo treatment in Mo deficient soils.

L. Phosphorus Deficiency

Next to N, P is probably the second most important growth limiting factor for plants growing in soils (Fox, 1979). Phosphorus deficiency is a major yield limiting factor in acid alfisols, oxisols, ultisols, and andepts (Clark, 1982). According to Sanchez and Salinas (1981), nearly all acid soils of tropical America are P deficient. An estimated 70% of these soils cause problems of Al toxicity and Ca or Mg deficiency and many have high P fixation capacities. Phosphorus fixation is one of the most important growth limiting factors for crops in Brazil (Fageria and Filho, 1987). Extractable P values (Mehlich No. 1) declined rapidly for 17 days before leveling off. The amount of P fixed ranged from 77 to 90% of that applied, depending upon the level added; hence, very high initial P applications are needed when such soils are first brought into cultivation. In acid soils, tolerance to high Al saturations is difficult to separate from tolerance to low levels of available P (Salinas and Sanchez, 1976), and liming soils to neutralize exchangeable Al greatly reduces the need for P fertilizers (Kamprath and Foy, 1985). For example, four times as much superphosphate was required on a soil of pH 5.1 and 49% Al saturation as on a limed soil at pH 6.1 with no Al to produce equal yields of alfalfa (Munns, 1965).

Surface layers of cultivated soils are usually fertilized with P, and band-placed P greatly stimulates root development in these zones, particularly when combined with NH_4^+-N. However, when the plow layer becomes dry, the plant needs to absorb P from deeper soil zones. Does P deficiency in the subsoils limit root growth? DeMiranda and Rowell (1987) found that liming subsoils improved plant growth, but yield was still limited by low P availability. Additions of P to the topsoil (presumably with adequate soil moisture) caused good growth regardless of subsoil acidity. Root growth increased in both layers and labeled ^{32}P absorbed in the topsoil was translocated to the roots in the subsoil. Thus, P was believed to inactivate root Al and allow roots to grow and take up additional P from the acid subsoil. Aluminum sensitive wheat cultivars were affected more by acid subsoil and low P availability than Al tolerant cultivars. Aluminum tolerant and sensitive cultivars had similar abilities to translocate P to roots in the subsoil, but sensitive cultivars could not attain growth rates equivalent to those of Al tolerant cultivars. Clark et al. (1988) concluded that Al tolerance in sorghum was not associated with tolerance to low P. However, there is evidence to the contrary for wheat and beans (Salinas and Sanchez, 1976), maize (Farina et al., 1982), and rice (Fageria, 1985).

McLaughlin and James (1989) studied the effects of subsurface Al on the growth and uptake of surface-applied P by wheat seedlings in a split root system consisting of sand in the upper part and a nutrient solution in the lower part. Root growth into the subsurface solution was increased by increasing P supply to surface roots in a sand culture—but only when the subsurface solution contained no Al. Increasing the P supply to surface roots had no effect on penetration of roots into the Al-containing solution. Aluminum tolerance in wheat was not related to the ability to translocate P to the point of Al injury.

In low P soils, root growth apparently takes preference over top growth (Classens and Folscher, 1988). Poor root growth of wheat plants, associated with low P, was characterized by higher root/top ratios than in plants supplied with adequate P. Roots also contained higher P concentrations than tops when grown with low P supply. Wheat seemed to require a certain level of P in roots before P was translocated to tops.

M. Silicon Deficiency

Highly weathered, humic, ferruginous latosols in Hawaii are low in soluble and total Si (Fox et al., 1967). Sugar cane (*Saccharum officinarium* L.) plants grown on these soils were also low in soluble and total Si, and yield responses to applied Si corresponded closely to concentrations of soluble Si in soils and in irrigation water. Silicon is not only beneficial to Si accumulating plants such as rice, horsetail (*Equisetum arvense* L.), and scouring rush (*Equisetum hayemale* L.) but also to nonaccumulators like tomato (*Lycopersicon esculentum* Mill.) (Takahashi and Miyake, 1976, 1982). Silicon is considered essential for sugar cane (Plucknett, 1972; Elgawad et al., 1982b) and for other plants having a high Si content, including barley, wheat, oat, bamboo (*Bambusa multiplex* [Lour.] Raeusch), sorghum, flax, grasses, sunflower, and garden beets (*Beta vulgaris* L.) (Cheng, 1982). Growth responses to added Si have been reported for cucumber (*Cucumus sativus* L.), soybean, fernbrake (*Pteridium aquilinum* [L.] Kuhn), barnyard grass (Takahashi and Miyake, 1982; Miyake and Takahashi, 1985) and wheat (Kundu et al., 1986). Yield responses of rice and sugar cane to calcium silicate slag on histosols of the Florida Everglades have also been attributed primarily to Si (Elgawad et al., 1982a,b; Snyder et al., 1986; Anderson et al., 1987).

Beneficial effects of Si have been attributed to increased P efficiency (Birch, 1953; Kundu et al., 1986), precipitation of Al and Fe (Clark and Gourley, 1988; Li et al., 1989), and alleviation of Mn toxicity by promoting a more uniform internal distribution of Mn, which prevents localized accumulations associated with necrotic spots on leaves (Cheng, 1982; Horiguchi, 1987, 1988; Galvez et al., 1989). Silicon stiffens plant stems and provides oxidative power to rice roots; this precipitates Fe and Mn on root surfaces and prevents excessive transport of these potentially toxic ions to plant tops (Cheng, 1982).

Takahashi and Miyake (1976) suggested that Si has a role as a micronutrient in the reproductive growth of various plant species as well as its role as a macronutrient; they reported Si deficiency in tomato and correction of the problem by adding Si to the nutrient solution. Silicon is reported to protect rice plants against grain discoloration while increasing grain yields by 34% on ultisols of Nigeria (Yamiguchi and Winslow, 1989). Clark and Gourley (1988) suggested that Si protected sorghum against Al and Mn toxicities on a Colombian oxisol. Silicon slightly reduced Al toxicity in some cotton cultivars in a nutrient solution at pH 4.5 (Li et al., 1989), but these beneficial effects were much less than those reported for sorghum (Galvez et al., 1989). Li et al. (1989) speculated that the beneficial effects attributed to Si by Galvez et al. (1989) were due to increased pH and precipitation of Al from the solution, rather than a direct effect of Si on Al toxicity. Li et al. (1989) also thought that the limited effect of Si in alleviating Al toxicity in their own experiment was due to the slow formation of Al-Si complexes.

Contrary to the findings of others, Jarvis and Jones (1987) reported that Si did not affect the growth of perennial ryegrass and white clover in a flowing nutrient solution. There was also no effect of Si on the absorption or transport of Mn.

Accumulation of Si in plants has been implicated in human health. Sangster et al. (1983) reported that four species of *Phalaris* accumulated Si in the form of sharp fibers which fall within the critical size range classified as carcinogenic. Seeds of these species are known contaminants of cereal crops in the Middle East, and such contamination has been correlated with a high incidence of esophageal cancer in northeastern Iran. Silicon contents were higher in heavily grazed than in ungrazed grasses in the Serengeti region of Africa. The accumulation of Si is believed to be a protective mechanism against grazing (McNaughton et al., 1985). These investigators, using plastic containers to prevent Si contamination, concluded that Si was not required for growth, but did promote flowering and increased leaf size and chlorophyll content.

In regions of high rainfall, a loss of the basic cations Ca, Mg, K, and Na, in addition to the resulting acidification of the soil, leads to the decomposition of clay minerals (aluminosilicates) and the leaching of soluble Si from the profile. Thus, oxisols and ultisols are lower in soluble

Si than less weathered soils. On such soils, plants having a high requirement for Si (tendency toward accumulation) might be expected to give growth responses to added Si. The beneficial effects of Si could result from supplying an essential nutrient or the complexation and detoxification of Al, Mn, and other ions. Silicon may be involved in the differential tolerances of plant genotypes to Al and Mn. The possible role of Si in the detoxification of acid soils and the adaptation of plants to such soils deserves further study.

N. Salinity

Salinity can suppress seed germination (Kpoghomon et al., 1986; Alwan et al., 1989; Shokohiford et al., 1989) and inhibit root growth indirectly through injury to plant tops (Pessarakli et al., 1989a,b). Salt stress severely decreases N metabolism and protein synthesis; injury is characterized by the accumulation of NH_4^+, NO_3^-, and free amino acids (Pessarakli et al., 1989a). Whether this results from protein degradation or the impairment of amino acid incorporation into proteins has not been determined. Sodium chloride directly reduces nitrate reductase activity in the roots of tomato and cucumber by inhibiting transport of NO_3^- from vacuole to cytosol (Martinez and Cerda, 1989). These effects of NaCl were accentuated by the presence of NH_4^+ as the N source. Postel (1990) estimated that crop yields are being reduced by salinity on 25 to 30% of the irrigated acreage in the United States.

III. Solving the Shallow Rooting Problem

A. Modifying the Soil

In most soils, conventional liming of the plow layer does not neutralize subsoil acidity. The most effective procedure for promoting root growth is to mix lime throughout the entire soil volume in which roots are desired. However, in most situations, this procedure is not economically feasible.

Various approaches have been tried to detoxify subsoil Al and thereby promote root development in deeper soil zones as protection against drought. Lime has been applied unmixed on the soil surface, mixed with the plow layer, and mixed with deeper layers by deep plowing, placed in a band on the plow sole, and injected (in dry or suspension form) behind subsoiling chisels. Another approach involves the application of gypsum ($CaSO_4 \cdot 2H_2O$) or phosphogypsum, a by-product of phosphoric acid manufacturing, to leach Ca to deeper zones of the profile where it replaces Al on the cation exchange complex (CEC). This procedure apparently works well in sandy soils or in oxisols (clay loams aggregated so as to behave like sands) where the Al displaced by Ca can be leached out of the root zone. However, if the displaced Al is not removed (as in finer textured soils), the gypsum treatment can lower soil pH and increase Al solubility and toxicity.

Another approach has been to apply heavy rates of lime and acid forming N fertilizers to induce the downward movement of Ca as $Ca(NO_3)_2$ and the displacement of Al. Lime has also been applied with ordinary superphosphate, which contains gypsum, to accomplish the same end—namely, the downward leaching of Ca. In some cases, deep placement of fertilizers, particularly P and K, has been used to increase root development in subsoils.

In the text to follow, selected references from the literature will be cited to show the results of these attempts to correct the shallow rooting problem by modifying the soil.

In coarse textured soils, such as sands and loosely aggregated oxisols, some downward movement of Ca from limestone has been reported. Bataglia et al. (1985) found that significant downward movement of Ca occurred when lime was applied for four successive years to a Brazilian oxisol. In other studies on oxisols, Ritchey et al. (1981, 1982) and Ritchey (1987) reported that applying limestone at 2500 kg Ca/ha, plus ordinary superphosphate, improved rooting to a depth of 75 cm, presumably by the downward movement of Ca and the displacement of Al. Root penetration was believed to be limited by the lack of Ca rather than Al toxicity.

In studies by Pearson et al. (1962) on two clay soils of Puerto Rico, high rates of acid-forming N fertilizers caused appreciable downward movement of basic cations in the profile; this movement increased with increased rates of surface-applied limestone and with increased residual acidity of the N source. Amounts of Ca and Mg moved from the soil surface into the 6 to 24 in. zone were equivalent to 5.9 tons of $CaCO_3$/acre. Although half of the Ca and Mg accumulating in the 6 to 24 in. zone was in exchangeable form, appreciable amounts were present as neutral salts; this may explain why the subsoil pH remained lower than expected.

Calcium carbonate mixed with acid subsoil produced large responses in the length of deep roots by some species (Simpson and Pinkerton, 1978). For example, lucerne shoots responded markedly (100 to 200% increase) to subsoil liming when subsoil moisture was available; however, when only surface water was supplied, the response was small (less than 30%). Messick (1982) applied 0, 4.48, 8.96, and 13.44 metric tons of dolomitic limestone per hectare, mixed with the 0 to 20 cm soil layer, to Goldsboro sandy loam (siliceous, thermic aquic Paleudult) and Tomotley loam. Three years after treatment, surface lime treatments had improved subsurface soil chemical properties. All lime treatments increased soil levels of exchangeable Ca and Mg and pH and decreased exchangeable Al in both soils at the 20 to 25 cm depth. Increases in exchangeable Mg were detected at 50 cm for the two highest lime treatments. Adams et al.

(1967a,b) concluded that acidity in the deeper soil zones of a Cecil soil (clayey, kaolinitic, thermic typic Hapludult) profile could be corrected by heavy applications of limestone and acid forming N fertilizers over a period of three to seven years. On a Coastal Bermuda grass sod, surface application of limestone did not increase soil pH below 15 cm after three years. Mixing limestone with the 0 to 15 cm layer increased soil pH only to the 30 cm depth with low N rates; however, at higher N rates, the effect was noted in the 45 to 60 cm zone.

Rechcigl et al. (1985c, 1988) concluded that surface soil application of lime was adequate for the establishment and productivity of alfalfa on an acid clay loam soil in Virginia. Roots penetrated the soil to a depth of 84 cm when the pH was only 4.1 and the level of exchangeable Al was 225 mg/kg. Alfalfa yields were related to root penetration and to increases in the N concentrations in plant tops. They concluded that reduced N_2 fixation was a major factor limiting root penetration. In another study (Rechcigl et al., 1987), treating soil in implanted bags with $Ca(OH)_2$ increased N_2 fixation four-fold in the surface limed soil and twenty-fold in the unlimed soil.

Long (1959) reported that top and root growth of several crops were more closely related to the effective moisture content and degree of aeration of the soil than to its fertility status. In a Lintonia soil, a subsoil pH of 4.7 did not hinder root development of alfalfa. Moschler et al. (1960) found that the beneficial effects of lime on alfalfa grown in an acid Tatum soil (subsoil pH 4.6) were not attributable to increases in the depth of root penetration. Beneficial effects of lime were thought to be due to the detoxification of Al and improved N metabolism.

After $8\frac{1}{2}$ years, an application of eight tons of lime/ha reduced soluble Al in the 30 to 45 and 60 to 75 cm layers of a Zambian ultisol to one-third and one-half, respectively, of that in the control, but lower rates of lime were ineffective (McKenzie et al., 1988). Placing lime at greater depths significantly increased the yields of Al sensitive wheat cultivars "Jupateco" and "Hornbill" but not of Al tolerant "Whydah." Wheat yield was highly correlated with soluble Al, particularly that at the 30 cm depth. Gypsum was less effective than lime in reducing soluble Al in the 15 to 30 cm soil layer.

Pearson et al. (1973) found that the growth of cotton roots in a Dothan subsoil (pH 4.6) was best when the entire mass of acid subsoil was mixed with limestone; however, even poorly mixed limestone, when applied at adequate total rates, increased the rooting depth of cotton by two-fold. Both radicle elongation and final rooting pattern were closely related to the percentage of the total subsoil mass that was limed. They estimated that the soil pH should be about 5.5, with 66% of the soil being neutralized, to obtain 50% of the maximum root growth rate. It was also estimated that root growth rate must be below 50% of maximum in order to reduce the yield of lint cotton by a significant amount. Pearson et al.

(1973) concluded that satisfactory mixing of lime was achieved when the mean distance between the neutalized masses was 5 cm or less. They suggested that injecting lime behind a subsoil opener and distribution of lime along the entire depth of the slit would improve root penetration if crops were planted directly over the slit each year.

Doss et al. (1979) studied the effects of combining subsoil chiseling and deep placement of lime on the root growth of cotton and corn on a soil having pH values between 4.1 and 4.8 before liming. All soil plots were chiseled to a depth of 45 cm and lime was incorporated to depths of 15, 30, and 45 cm. Deep placement of lime increased rooting depth, soil water extraction, plant height, and yields of cotton and corn. Incorporation of lime to a 30 cm depth was sufficient to produce satisfactory root growth for high yields of both crops. Cotton was more sensitive to the acid soil than corn.

According to Brooke et al. (1989), liming and/or deep ripping to break compact layers increased the yields of a wide range of crops. Wheat, oat, triticale (*X. tritosecale*), safflower (*Carthamus tinctorius* L.), chick pea, (*Cicer arietinum* L.), barley, rapeseed (*Brassica napus* L.), and field pea (*Pisum sativum*) responded to lime but lupine (*Lupinus* sp.) did not. Safflower and chick pea were particularly sensitive to the acid soil. Deep ripping without lime decreased yields. Simpson et al. (1979) grew alfalfa in 85 cm columns containing limed, fertilized surface soil, underlain by a low Ca, acid subsoil. Soil pH was the most important factor controlling root elongation; at low rates, $CaCO_3$ was more effective than $MgCO_3$. Surface drought reduced shoot growth, but a response to subsoil liming was observed as long as the subsoil moisture was adequate. Short term growth under surface drought was correlated with the total length of deep roots (greater than 60 cm). In studies by Mallett et al. (1987), maize roots were restricted to the depth to which lime was incorporated in a Normadien soil (Plinthudult) at Geluksburg, South Africa. They concluded that lime should be incorporated to a minimum depth of 55 cm (21.8 in.) to allow crops access to water and avoid drought. Deep band placement of suspension and dry lime at a depth of 20 to 40 cm (8–14 in.) into an acid subsoil raised soil pH and increased cotton yields (Ebelhar and Tupper, 1985). Benefits were greatest during dry years when rooting depths were greatest. In this instance, lime was applied behind a parabolic super chisel shank equipped with rolling disc hippers (bed formers). Chiseling and hipping without lime brought acid subsoil to the surface and decreased yields.

Adeoye and Singh (1985) noted that deep mixing of soil, with or without lime, increased the yield of maize, sorghum, and peanut when compared with shallow mixing on an acid, sandy soil. Incorporation of lime by deep moldboard plowing and incorporation of lime bands to a depth of 0.7 meters with a Wye double digger were superior to the conventional moldboard plow incorporation of lime (Farina and

Channon, 1980). In that study, lime response varied from $1400\,kg\,ha^{-1}$ with severe water stress to $400\,kg\,ha^{-1}$ during a high rainfall season. Yield responses were related to rooting depth and a change in the root system configuration. Placing lime in deep, vertical bands was as effective as deep plowing or double digging. Tupper et al. (1987) also found that deep band placement of dry lime increased soybean yield and produced economic returns which more than offset the added costs of the treatment.

In some experiments, deep placement of fertilizers has improved plant growth. Lu and Edwards (1985, 1986) concluded that some form of deep tillage is required on the coarse textured, sandy soils of the southeastern United States in order to obtain the maximum yield of wheat. They used slit tillage, plus chemical injections, to break plowpans and promote root development in subsoils. Injecting calcium nitrate, calcium phosphate, ammonium phosphate, and dolomitic limestone into a subsoil at pH 4.3 increased sorghum heights by 15%, but ammonium nitrate decreased plant height by 15%. In the check treatment, 29% of the roots were in the subsoil; the corresponding value for ammonium phosphate was 49%. Injecting P containing fertilizers into the subsoil increased concentrations of P in plant leaves and roots. Soil pH and exchangeable Al levels were decreased by dolomitic limestone only within 12 mm of the point of injection.

Incorporating P and lime by deep plowing increased levels of P, Mg, and Ca in the 18 to 38 and 38 to 51 cm depths and created a more favorable rooting environment (Cassel, 1980). During years of moderate to severe drought, subsoil loosening increased yields of spring-planted crops on 16 sites of well-drained, light to medium textured soils in an arable rotation (Marks and Soane, 1987). This response was associated with deeper rooting and improved water extraction from subsoils. In only one site was subsoil placement of P and K better than fertilizing the surface soil. Deep placement of monocalcium phosphate increased the P content and yield of drought-affected plants, but yields were increased further when lime and B were placed with P in deep bands (Simpson and Lipsett, 1973). These effects were not observed in plants grown with moist surface soils. These investigators noted that Australian soils show a large decrease in P with increasing depth and suggested that drought may be due to the interactions of nutritional factors and soil moisture stress rather than simple moisture deficits.

McEwen and Johnson (1979) found that subsoiling alone increased yields of wheat by 21%, barley 24%, and sugar beet by 11%. Increasing P and K in the subsoil increased the mean yields of potatoes (*Solanum tuberosum* L.) by 16%, barley by an additional 20% and sugar beet by 4% over those obtained by subsoiling alone. Wheat yield was not further benefited by the addition of P and K in subsoils. Kapoor et al. (1982) noted that, when fertilizers were added to the 0 to 15, 15 to 30, and 30 to 45 cm soil layers, N in all three zones contributed to N uptake but P was available only in the upper soil layers. Potassium in all three layers was available to maize, but only K in the upper layers influenced K absorption by wheat.

Rechcigl et al. (1985a,b) reported that plowsole application of dolomitic limestone (22,400 kg ha^{-1}) on Davidson (kaolinitic, thermic rhodic Paleudult) clay loam reduced exchangeable Al and increased pH, root length, and alfalfa yields, but the addition of P and K to the plowsole did not affect yields. Growth of wheat roots in a subsurface nutrient solution (sand culture above) was increased by increasing P supply to the roots, but only when Al was absent from the solution (McLaughlin and James, 1989). These results did not support the conclusion that P absorbed from a surface medium can neutralize subsurface Al. The authors suggested that P absorbed by roots growing in the surface medium and translocated downward can only ameliorate subsurface Al outside the plant roots by the leakage of $H_2PO_4^-$. Aluminum tolerance of wheat cultivars was not related to the ability to translocate P to the site of root injury. Potassium (Rb86) applied to subsoils was transferred from alfalfa to maize grown in a sand culture (surface medium) within 40 days and to maize within 10 days (Habben and Blevins, 1989). Manson and Fey (1989) found evidence that the addition of low levels of Na (less than 2 meq kg^{-1}) to low Na soils may alleviate acid soil toxicity. Additions of NaCl decreased the activity of Al by 10-fold and increased rooting in acid subsoil. Comparable additions of KCl did not affect root growth. These investigators suggested that Na applied to surface soils as Na_2CO_3 or Na_3PO_4 would increase subsoil pH and improve root development. In this connection, $NaNO_3$, applied in a band at planting time, is reportedly more effective than NH_4NO_3 in promoting early growth of cotton (G.R. Burns, pers. comm.). The difference could reflect the fact that the former is residually basic and the latter residually acidic.

Gypsum ($CaSO_4 \cdot 2H_2O$) has been used to alleviate both physical and chemical growth-limiting factors for roots. Radcliffe et al. (1986) reported that gypsum increased subsoil root activities which, in turn, reduced mechanical impedance. Sumner and Carter (1987) used deep liming and surface gypsum treatments on alfalfa grown on Appling (clayey, kaolinite, thermic typic Hapludult) soil in Georgia. Yield responses to both were highly significant and were attributed to a reduction in subsoil Al and an increase in Ca. Root examination showed that yield responses were due to increased root growth in subsoils and increased resistance to drought. However, the increased root proliferation caused a P deficiency in plants, despite a high overall soil P test. In this study, root length was a better index of root surface than root mass. Gillman and Sumner (1987) noted that Ca from gypsum moved rapidly through soil profiles, causing Mg to be displaced from exchange sites and lost by leaching; hence, Mg fertilization may be needed. Liming an acid soil to a depth of one meter neutralized Al, increased levels of soluble Ca, stimulated root development in subsoils, and increased alfalfa yield by 50% (Sumner et al., 1986). By comparison, surface application of gypsum reduced the activity of Al (to a lesser degree than did liming), increased the activity of Ca in saturated extracts of subsoils, and increased alfalfa yields by 25%.

Farina and Channon (1988) concluded that surface incorporation of gypsum ($10\,Mg\,ha^{-1}$) is an economical, viable approach to the amelioration of acid subsoils of South Africa. Initial costs of the treatment were recovered within three years and benefits were expected to persist for many years thereafter. Four years after the surface application of gypsum, the treated plots had decreased levels of exchangeable Al and increased exchangeable Ca, Mg, and SO_4-S in the subsoil, and grain yields were increased by $13.4\,Mg\,ha^{-1}$. Phosphogypsum, the form of gypsum most readily available in South Africa, also supplies P in a form available to maize. Gypsum was believed to provide a self-liming effect by the formation of an AlOH-SO_4 mineral. After 1990, the use of phosphogypsum for agricultural purposes may be regulated or prohibited because this waste product of phosphoric acid production contains uranium and radium which may give rise to hazardous emissions of radioactive radon gas [EPA Environmental Standards Branch, Criteria and Standards Division (ANR-460), Office of Radiation Safety, EPA, Washington, D.C. 20460].

Ritchey et al. (1981) observed that maize grown on a dark red, Brazilian latosol and fertilized with triple superphosphate (TSP) showed wilting, but plants grown with ordinary superphosphate (OSP) did not. Roots of the OSP treated plants extracted water from at least the 120 cm depth, but no roots of TSP treated plants were found below the 45 cm depth where the Al saturation of soil CEC was greater than 60%. The OSP treatment increased subsoil pH and Ca and Mg levels and decreased Al saturation to the 75 to 90 cm depths. The beneficial effects were attributed to the gypsum content of OSP (50%). Later studies showed that broadcast applications of $5000\,kg\,ha^{-1}$ $CaSO_4 \cdot 2H_2O$ or a $2500\,kg\,ha^{-1}$ band placed at 80 cm intervals reduced wilting and increased Ca and Mg in the subsoil. Surface applications of gypsum (1300 to $3300\,kg\,ha^{-1}$) caused Ca to leach to a depth of at least 120 cm and significantly reduced Al saturation of the subsoil. The root restricting factor in this soil was believed to be Al (0.2 to 0.6 meq/100 g) which was displaced in the subsoil by Ca in the gypsum. The same investigators reported that some clayey B horizons of red yellow latosols in Brazil had pH values of 5.7 or higher and no measureable exchangeable Al, but root growth in these horizons was still severely restricted. The problem was attributed to Ca deficiency, rather than Al toxicity.

In studies by Black and Cameron (1984) on the reconstructed profile of an acid (pH 5.39) New Zealand soil, liming the A horizon did not alter pH, exchangeable Al, Al/Ca ratio, or root yield in the B horizon. Liming the B horizon increased pH, decreased Al and Al/Ca ratios, and increased alfalfa root growth in that horizon. Surface applied gypsum (13.7 tn. ha^{-1}) increased exchangeable Ca but decreased pH and only slightly decreased the Al/Ca ratio in the B horizon and did not increase alfalfa root yields. Heavy liming of the surface soil (pH 7.8) could cause problems of micronutrient availability and was, therefore, not considered a practical solution to the problem of subsoil acidity. Fernandez and Blue (1984) obtained yield responses in maize by applying lime-gypsum to columns of Dothan sandy loam in Florida. Yield responses were directly related to increased pH, reduced Al saturation, and improved base status in surface and subsoil. Soon (1988) found that $CaCO_3$ increased the yield of barley and rice but $CaSO_4$ did not; gypsum increased Al concentration in the soil solution.

Beneficial effects of phosphogypsum (PG) have been attributed, at least in part, to ion pairing of Al with F and SO_4 and the alleviation of Al toxicity (Alva and Sumner, 1988; O'Brien and Sumner, 1988; Alva et al., 1989). In nutrient solutions containing 40 μM Al (1 ppm), the addition of $CaSO_4 \cdot 2H_2O$ (2.07 g/l) decreased the predicted activity of Al^{3+} from 36 to 11 μM and increased that of $AlSO_4^+$ from 33 to 80 μM while increasing the lengths of soybean roots four-fold. When soil was treated with 5 Mg ha^{-1} of PG, 99 and 0.6% of the total Al in the soil leachate was complexed with F and SO_4, respectively, but only 0.3% was in the Al^{3+} form. With 10 Mg ha^{-1} of PG, 10% of the total Al in soil leachate was in the Al^{3+} form, but the activity of Al was only 0.11 ppm (4.4 μM). Increases in the ionic strength of the solution, as well as ion pairing, could have decreased the activity of Al. In column leaching experiments, O'Brien and Sumner (1988) found that surface application of PG reduced subsoil acidity, as evidenced by reduced levels of exchangeable Al and increased Ca in subsoils. They suggested that Al was detoxified as a solid phase, possibly alunite.

Fluidized bed combustion residue from a calcitic limestone source, a by-product of scrubbing SO_2 from coal-fired boilers, has been an effective substitute for agricultural limestone (Edwards et al., 1985a,b).

Another approach to the amelioration of acid subsoils is the incorporation of green manure crops, animal manure, or other organic matter which may provide chelating ligands for the detoxification of Al and the prevention of P fixation by Fe and Al. Hue et al. (1986) demonstrated that a given concentration of soluble Al in uncultivated soils is less toxic than in cultivated soils, probably because Al in the former is chelated in soluble but non-toxic forms. A green manure crop of sesbania increased yield and P efficiency in maize, both in the absence and presence of added P fertilizers. Chelation may be a means of mobilizing soil Al and removing it from the root zone. Hern et al. (1982) showed that a synthetic chelating agent, EDTA, mobilized significant amounts of Al to a depth of 60 cm in leached soil columns. Also working with soil columns,

Wright et al. (1985) obtained maximum root growth and rooting depth with manure treatments. In the same experiment, lime and gypsum treatments were no better than the check.

Correction of salinity generally involves treating a soil with gypsum and flushing with less saline water to remove excess salts. Amendments of leaf litter and manure reportedly improve seed germination, plant germination, and plant growth, mainly by inhibiting excess flow of Na into plant cells through the ion holding medium (Shokohiford et al., 1989). Stassart (1987) reported that the addition of K to a medium of low salt roots (100 m*M* NaCl with or without to 10 m*M* $CaCl_2$) decreased primary accumulation of Na and did not affect Na accumulation but reduced Na mobility. Potassium plus Ca reduced Na accumulation in high salt roots, but K increased Na flux in the xylem.

B. Using Stress Tolerant Plants

When modifying the soil by conventional means is not economically feasible, a reasonable alternative or supplemental approach to the shallow rooting problem is to use plant genotypes having greater tolerance to root-limiting factors. Plant species, and genotypes within species, vary widely in tolerance to mineral stress factors of toxicity and deficiency (Clark, 1982, 1984; Foy, 1983, 1984; Foy et al., 1988; Little, 1988). Stress tolerant genotypes are being identified and bred worldwide to solve some of the more difficult problems of soil fertility, such as subsoil acidity, salinity, low plant availability of Fe in calcareous soils, and low P availability in acid soils. Plant genotypes are also being sought to improve fertilizer efficiency on "good soils." This approach of "tailoring the plant to fit the soil" is compatible with the newly emphasized low input sustainable agriculture (LISA). For example, Al tolerant plants may be more drought tolerant and require lower inputs of lime and P fertilizer than less tolerant genotypes. Nitrogen efficient genotypes might save on fertilizer costs and reduce the probability of groundwater pollution by nitrates. Because of its tolerance to Al and other toxic ions, Bermuda grass is recommended for the reclamation of acid coal mine spoils before planting less tolerant but more profitable crops (Taylor et al., 1989).

Differential Al tolerance has been found among many plant species (Neves et al., 1982a,b; Bueno et al., 1985; Gallo et al., 1986; Saneoka et al., 1986; Bilski, 1987; Taylor et al., 1989). Within species, differential Al tolerance has been identified or developed in rice (Fageria et al., 1988a,b; Hai et al., 1989; Jan and Pettersson, 1989), sorghum (Duncan, 1981, 1984, 1988; Furlani and Clark, 1981; Ramirez et al., 1986; Gourley, 1987; Duncan et al., 1989; Hill et al., 1989); wheat (Muzilli, 1984; Lafever, 1988; Little, 1988; Ritchey et al., 1988a,b; Berzonsky and Kimber, 1989); soybean (Floyd and Mookherji, 1987; Klotz and Horst, 1988); red clover (Baligar et al., 1987), annual ryegrass (*Lolium multiflorum* Lam.)

(Rengel and Robinson, 1989), cotton (Kennedy et al., 1986), potato (Wersuhn et al., 1988), *Nicotiana plumbaginifolia* (Meredith et al., 1988), maize (Ramirez and Berensel, 1984), alfalfa (Devine et al., 1976; Simpson et al., 1977; Simpson and Pinkerton, 1978; Bouton et al., 1986; Baligar et al., 1989; Campbell et al., 1989; Parrot and Bouton, 1990), amaranthus (*Amaranthus* sp.) (Foy and Campbell, 1984; Campbell and Foy, 1987), leucaena (Oakes and Foy, 1984) and Old World bluestems (*Bothriochloa intermedia* [R. Br.], *A. Camus*) (Foy et al., 1989).

Differential tolerance to low pH has been reported in white clover (*Trifolium repens* L.) (Jarvis, 1984) and soybean (Reddy and Dunn, 1987). Poinsettia (*Euphorbia pulcherrina* Wilde. *Ex. Kl.*) cultivars differ in Mo requirement (Cox, 1988). Plant species and genotypes within species differ in efficiencies for using Ca and Mg (Clark, 1982, 1984; Fageria et al., 1989). Differences in P efficiency are known among plant species (Fox, 1979; Adams and Odom, 1985; Fox et al., 1986; Fist et al., 1987; Foehse et al., 1988; Amann and, Amberger, 1989; Fageria and Baligar, 1989) and among genotypes of wheat (Camargo and Felicio, 1987; Romer et al., 1989), tomato (Coltman, 1987; Fageria et al., 1988b), and white clover (Caradus and Snaydon, 1986).

Differential Mn tolerance occurs among species of fungi (Babich and Stotsky, 1986; Smith et al., 1987), among species of higher plants (Bilski, 1988; Fageria et al., 1989), and within many plant species, including soybean (Mascarenhas et al., 1985; Reddy, 1987; Reddy and Dunn, 1987; Foy et al., 1988).

Differences in tolerance to salinity have been reported among genotypes within snap beans (Pessarakli et al., 1989a,b), maize (Alwan et al., 1989), and other species. For additional information on salinity, see a review by Stassart (1987).

In addition to providing for crop production under current conditions, stress tolerant plant genotypes may improve the soil, both physically and chemically, so that plants having more exacting growth requirements (and possibly higher yield and greater profitability) may be grown later. For example, Al tolerant plants such as limpograss, *Hemarthia altissima* (*Poir*) *Staph* and C.E. Hubb (Foy and Oakes, 1984), may create channels that would benefit crops to follow. In addition, root growth of tolerant plants in an acidic zone containing high concentrations of Al could produce decomposition products that could detoxify Al and prevent P fixation by Al and Fe (Traina et al., 1986; Helal and Dressler, 1989).

Stress tolerant, compatible, legume-rhizobial combinations are needed for adaptation in the acid, infertile soils used for the production of forage in a low input system. There appears to be ample opportunity to select for stress tolerance in host plant species, rhizobial strains, and combinations of the two.

Rhizobial species and strains differ markedly in tolerance to low pH and toxicities of Al and Mn in tropical soils (Jadi, 1982). *Rhizobium lupini* tolerated a pH as low as 3.15. Cowpea rhizobia did not survive below pH 5.0 (Sundram, 1979). One strain of *R. phaseoli*, S-442, grew well at pH 4.0 while strain S-422 did not (Amara et al., 1981). Forty strains of cowpea rhizobia on two cultivars of mung bean varied in ability to nodulate. A few strains were very sensitive and did not nodulate at pH 5.0, while some supported good growth of mung bean (Munns, 1979). Keyser et al. (1979) reported similar variations in tolerance of cowpea rhizobia to acid soils.

Paulino et al. (1987) found that low pH (5.2) decreased nodule numbers and acetylene reduction in a *Rhizobium leguminosarum-Pisum sativum* association and that Al further depressed these parameters. Significant bacterial strain effects on nodulation were noted in the presence of different Al concentrations. Nitrogenase activity was decreased by low pH and further by Al. Concentrations of 120 μM Al stopped nitrogenase activity but did not interrupt nodule formation. In this association, Al was more harmful than low pH. Rhizobial strains used to inoculate *P. vulgaris* differed in tolerance to low pH and Al in Kenyan soils (Karanja and Wood, 1988).

Strains of *R. japonicum* varied in tolerance to 50 μM Al, low pH (4.5), and low P (5 to 10 μM) (Keyser and Munns, 1979). Aluminum was judged more detrimental to rhizobial growth than low pH, low P, or low Ca in acid soils of the tropics. Jadi (1982) showed that strains of rhizobia varied in tolerance to Al at pH 5.0 and to low P (10 μm) in a solution culture. *Rhizobium japonicum* strains were tolerant to 100 μM Al. Some strains of *R. phaseoli* also tolerated 100 μM Al, but others were inhibited at 50 μM Al. In general, the slow growing genera of rhizobia were more tolerant to acidity and Al than fast growing strains (Graham and Parker, 1964).

Lindstrom and Myllyniemi (1987) studied 19 strains of *R. trifolii* isolates from Finland in an acidic laboratory medium. The lowest pH at which growth occurred (critical pH) was between 4.7 and 4.9. Five strains tested for tolerance to Al (50 to 150 μM) and Mn (200 to 730 μM) also showed effects of these ions at the critical pH but not at higher pH levels. At pH 4.1, the best strain tested improved plant growth 14-fold over that of the check treatment. Others have determined the critical pH of *R. trifolii* to be 4.6 to 5.2 (Bromfield and Jones, 1980), 3.0 to 4.5 (Hagedorn and Caldwell, 1981) and 4.2 to 4.6 (Thornton and Davy, 1983). There was no correlation between symbiotic performance of the strains in acid clay soil in the laboratory and in an acid laboratory medium, but the most tolerant strain was isolated from the most acidic site in the field. Hence, the pH of the isolation site was thought useful in predicting symbiotic performance in very acidic soils.

Aluminum concentrations of 22.5 μM differentially affected the nodA genes in three strains of *R. leguminosarum biovar trifolii* over a range of pH values, and Ca reduced the inhibitory effects of Al. It was concluded that poor induction of the expression of nodA genes in *R. leguminosarum biovar trifolii* contributed to the acid sensitivity of the infection process (Richardson et al., 1988a,b).

Chong et al. (1987) reported that peanut and cowpea were more tolerant to an acid (pH 3.9), high Al soil than mung bean (*Vigna radiata* [L.] Wilczek) and pigeon pea (*Cajanus cajan* [L.] Huth). Growth, nodulation, and N_2 fixation were higher in cowpea than in peanut and other legumes at pH 3.9.

Cooper et al. (1986) found that marsh trefoil (*Lotus peduncaulatus* Cav.) nodulated satisfactorily at a pH of 4.5 and showed a preference for *R. loti* in mixed inocula. White clover failed to nodulate at pH 4.5 in the absence of Al and at 5.5 in the presence of 50 μM Al. The toxicity of Al observed in *R. trifolii* when the Al was partly neutralized was attributed to polyhydroxy forms of Al. These authors suggested that the sensitivity of the nodulation process in white clover might be partially overcome by selecting Al tolerant strains of the rhizobia. Stress tolerant strains of rhizobia have been used to nodulate cowpea (Keyser and Munns, 1979; Hartel and Bouton, 1989) and for establishment of subclover on acid soils (Thornton and Davy, 1983). Cooper et al. (1986) suggested transferring the symbiotic plasmid or nodulation genes from *R. trifolii* to another fast growing *Rhizobium* species which already has acid soil Al tolerance. Because of its high tolerance to these stress factors, *R. loti* was regarded as an appropriate recipient species. In further studies, Wood and Cooper (1988) determined that in the absence of Al, the lowest pH for multiplication of *R. trifolii* was 4.8 to 5.0 for strain HP3 and 4.4 to 4.6 for strain BELI 192. Aluminum at a concentration of 50 μM (pH 4.6 to 5.6) inhibited multiplication in both strains, despite the precipitation of Al at pH values above 4.6. After cells had been exposed to Al for 14 days at pH 4.5 and 5.6, additions of 10 to 100 μM P did not remove the inhibitory effect of Al, but transfer to a no-Al medium allowed multiplication to resume. It was concluded that the toxic form of Al was still unknown.

In a growth chamber study, Hartel and Bouton (1989) found that an alfalfa strain selected for tolerance to acid soil and inoculated with a more acid tolerant strain of *R. meliloti* produced greater top growth, nodule numbers and weights, and acetylene reduction values in an unlimed soil at pH 4.6 than did the same alfalfa strain inoculated with a relatively acid sensitive rhizobial strain or an acid soil sensitive alfalfa strain inoculated with either a relatively acid tolerant or sensitive strain of *R. meliloti*.

Species and strains of rhizobia also differ in tolerance to high levels of Mn (Keyser and Munns, 1979). Some strains tolerate as much as 320 ppm Mn in the growth medium. Dobereiner (1966) found that strain 413 of *R. phaseoli* was more sensitive to 400 ppm Mn than strain 411. Franco and Dobereiner (1971) suggested that the processes of nodulation and N_2 fixation were more sensitive to Mn than were host plants.

Another approach to the problem of limited root development would be to identify and use adapted strains of soil fungi and bacteria to regulate the uptake of essential and toxic elements and water by plants (Muchovej et al., 1989). For an excellent review on the subject, see Kucey et al. (1989). Mycorrhizae are considered essential for the growth of some plants on low P soils of the tropics (Yost and Fox, 1979; Fox, 1981). Mechanisms by which mycorrhizae could increase the use of P by plants include the following: (1) absorption from sources unavailable to uninfected plants; (2) alteration of plant growth so that the plant produces a larger root system or alters enzymes involved in P absorption; and (3) extension of the P depletion zone away from the root (Kucey et al., 1989). The evidence to date indicates that mycorrhizal plants do not use unavailable (insoluble) P sources; instead, they use available solution P more efficiently than uninfected plants. Apparently, these fungi can use P from soil solutions that are too dilute for use by the host plant alone. Hence, properly adapted strains of mycorrhizae could replace part of the P fertilizer requirement for some crops. The VA mycorrhizal *Glomuus fasciculatus* can substitute for up to 56 ppm P in greenhouse grown Troyer citrange (rootstock of *Citrus sinensis* [L.] Osb. *X. Poncirus trifoliata* [L.] Raf. cross) and up to 278 ppm P in the Brazilian sour orange, *Citrus aurantium* L., (Menge et al., 1978). Some free living fungi and bacteria can solubilize P and make it available to higher plants. Furthermore, the beneficial effects of these organisms in improving P use are additive to those of mycorrhizae (Kucey et al., 1989). Selected strains of such organisms could be combined with adapted strains of mycorrhizae, Al tolerant host plants, and raw rock phosphate for use on acid soils in a low input system of agriculture.

According to the review by Kucey et al. (1989), the current interest in P solubilizing bacteria and fungi for the in situ dissolution of rock phosphate in soils has led to the application for several patents governing the biological release of P, but such systems have not yet been fully evaluated under field conditions.

In addition to increasing the uptake of P and water by roots, mycorrhizae may protect plants against excesses of toxic elements. For example, mycorrhizae reportedly decreased Al toxicity in pitch pine (*Pinus rigida* Mill) (Cumming and Weinstein, 1990), Norway spruce (Wilkins and Hodson, 1989) and switchgrass (Koslowsky and Boermer, 1989). In contrast to the reports above, Jones et al. (1986) reported that mycorrhizae reduced Ca uptake and increased the Al sensitivity of paper birch (*Betula papyrifera* Marsh) and jack pine (*Pinus banksiana* Lamb). Arines et al. (1989) reported that mycorrhizae decreased Mn concentrations in shoots and roots of plants in three acid soils.

IV. Challenges for Future Research

1. Develop economical methods for incorporating lime into acid, Al-toxic subsoils.

2. Assess the role of mechanical impedance in limiting root penetration of subsoils. In which physiographic regions and which soil associations is this a major problem? Does it occur in combination with Al toxicity?

3. Develop methods for keeping subsoil channels open after ripping. Should vertical mulching (blowing chopped crop residues into channels behind a subsoil chisel) be revived? Such placement of organic materials could increase the activities of soil microorganisms and earthworms and increase subsoil penetration by roots. In the process of decomposing, such organic matter could also provide ligands which complex and detoxify Al and prevent P fixation.

4. Determine the roles of Al and Mn toxicities in limiting root growth. In what soils and in what crops are these factors harmful? Do Al and Mn toxicities occur in the same soil for a given crop? For different crops? (Total Mn levels are higher in the Gulf coastal plain and in the Piedmont than in the Atlantic coastal plain.)

5. Examine the role of absolute Ca deficiency and/or Al × Ca interactions in regulating root growth in acid subsoils. Separate truly Ca deficient soils from Al affected soils.

6. Test the hypothesis that deep placement of P (and other fertilizer nutrients) is profitable during dry seasons.

7. Evaluate Si and F as potential detoxifiers of Al in acid subsoils and within plants. Relate the results to the use of by-products containing these elements. Are ultisols and oxisols deficient in Si for some plants?

8. Develop Al and Mn tolerant, drought tolerant, P efficient, and N efficient plant genotypes and legume-rhizobial combinations for use on marginal soils in a low input sustainable system of agriculture.

9. Identify rhizobial strain, host plant genotype, × acid soil factor interactions affecting root growth.

10. Determine the economic value of stress tolerant plant genotypes. For example, evaluate Al tolerance in terms of drought avoidance and savings in costs of irrigation, lime, and fertilizers. Could root penetration by an Al tolerant plant reduce the need for subsoiling by creating and maintaining root channels for subsequent crops?

11. Identify or develop stress tolerant strains of mycorrhizae for greater efficiency in the use of P and water and possible protection of host plants against toxic factors such as Al.

12. Re-assess the value of rock phosphate and P solubilizing soil microorganisms used in combination with Al tolerant plant genotypes in a low input system of agriculture.

13. Re-examine the value of winter cover crops and green manure crops in providing organic ligands which complex and detoxify Al and reduce P fixation by Fe and Al in acid soils.

14. Determine the physiological and biochemical mechanisms by which plants tolerate stresses such as Al and Mn toxicities. This involves characterizing the root-soil interface as well as the internal nature of plants.

15. Identify and characterize the genes controlling plant tolerances to Al and other stress factors. (Identification of gene products—proteins which might act to detoxify Al or offset other stresses.)

16. Resolve the question of whether or not $Al(OH)^{2+}$, $Al(OH)_2^{+}$, and polymeric hydroxy Al forms are toxic to plants.

17. Modify the GEOCHEM procedure for calculating Al activity to allow for water soluble but non-toxic Al-organic complexes in soil solutions.

18. Develop a soil test that will predict the occurrence and intensity of Al toxicity for all soils and all crops.

V. Summary

Perhaps the most important chemical limitation to root penetration and proliferation is that imposed by strongly acidic subsoils having pH values of 5.0 or below. In this situation, shallow rooting causes greater susceptibility to drought and less effective use of subsoil nutrients. The soil acidity complex may include toxicities of Al, Mn, and other elements, deficiencies of unavailabilities of Ca, Mg, P, Mo, and Si, and various interactions among these factors. The situation is further complicated by interactions of these factors with drought and particularly with plant genotypes. Aluminum toxicity is probably the most important growth-limiting factor in a majority of acid subsoils.

The shallow rooting problem can be attacked by modifying the soil, by using stress tolerant plant genotypes, or a combination of the two approaches. Soil modification practices include applying lime unmixed on the soil surface, mixed with the plow layer, mixed with deeper soil zones by deep tillage, placed in a band on the plow sole, and injected (dry or in suspension) behind subsoil chisels. In coarse textured soils, applying gypsum-containing by-products or a mixture of high lime and high N fertilizer treatments to the soil surface can sometimes leach Ca into the subsoil and displace toxic Al from the root zone. Although these procedures often improve root development, their economic feasibilities vary widely. The most effective procedure is to mix lime throughout the desired rooting zone. The second approach to the shallow rooting problem, using stress tolerant plant genotypes, is still in its infancy but has great potential, particularly in low input sustainable agriculture.

Challenges for future research on the rooting problem include: developing economical methods for liming acid, Al-toxic subsoils; separating

Al-toxic subsoils from those in which mechanical impedance limits root development; re-evaluating and perfecting vertical mulching procedures which will keep subsoil channels open; separating Al-toxic and Mn-toxic soils; separating truly Ca deficient soils from Al-affected soils; identifying the soils in which deep placement of P or other fertilizer increases root growth; evaluating Si, F, and SO_4 as detoxifiers of Al in acid soils and in plants; developing Al and Mn tolerant, P efficient, and N efficient plant genotypes for marginal soils; identifying rhizobial strain, host plant, and acid soil factor interactions affecting root growth; determining the economic value of stress tolerant plant genotypes; identifying or developing stress tolerant strains of mycorrhizae for greater efficiencies in the use of P and water and for possible protection of host plants against toxic factors such as Al; re-assessing the value of rock phosphate and P solubilizing soil microorganisms used in combination with stress tolerant genotypes for low input agriculture; re-examining the value of winter cover crops and green manure crops in providing organic ligands which complex and detoxify Al and reduce fixation of P by Fe and Al in acid soils; identifying the genes controlling stress tolerances in plants; determining the physiological and biochemical mechanisms by which plants tolerate stresses such as Al and Mn toxicities; identifying more precisely the toxic forms of Al in acid soils; modifying the GEOCHEM procedure for estimating Al activity to allow for water soluble but non-toxic Al-organic complexes in soil solutions; and developing a universal soil test for predicting the occurrence and severity of Al toxicity in acid subsoils.

VI. Acknowledgments

I thank T.E. Devine and R.R. Duncan for reviewing the manuscript.

References

Adams, F., and J.B. Henderson. 1962. Magnesium availability as affected by deficient and adequate levels of potassium and lime. *Soil Sci. Soc. Am. Proc.* 26:65–68.

Adams, F., and Z.F. Lund. 1966. Effect of chemical activity of soil solution aluminum on cotton root penetration of acid subsoil. *Soil Sci.* 101:193–198.

Adams, F., and B.L. Moore. 1983. Chemical factors affecting root growth in subsoil horizons of coastal plain soils. *Soil Sci. Soc. Am. J.* 47:99–102.

Adams, F., and R.W. Pearson. 1967. Crop response to lime in the southern United States and Puerto Rico. *In* R.W. Pearson and F. Adams (Eds.). *Soil Acidity and Liming*. Am. Soc. Agron., Madison, Wisconsin, (Agronomy 12) 161:206.

Adams, F., R.W. Pearson, and B.D. Doss. 1967a. Relative effects of acid subsoils on cotton yields in field experiments and on cotton roots in growth chamber experiments. *Agron. J.* 59:453–456.

Adams, J.F., and J.W. Odom. 1985. Effects of pH and phosphorus rate on soil solution phosphorus availability. *Soil Sci.* 140:202–205.

Adams, W.E., R.W. Pearson, W.A. Jackson, and R.A. McCreery. 1967b. Influence of limestone and nitrogen on soil pH and coastal Bermuda grass yield. *Agron. J.* 59:450–453.

Adeoye, K.B., and L. Singh. 1985. The effect of bulk application of lime under two tillage depths on soil pH and crop yield. *Plant Soil* 85:295–297.

Alam, S.M., and W.A. Adams. 1980. Effects of aluminum upon the growth and nutrition composition of oats. *Pak. J. Sci. Ind. Res.* 23:130–135.

Alva, A.K., and M.E. Sumner. 1988. Effects of phosphogypsum or calcium sulfate on aluminon reactive aluminum in solutions at varying pH. *Commun. Soil Sci. Plant Anal.* 19:1715–1730.

Alva, A.K., G.J. Gascho, and Y. Guang. 1989. Gypsum material effects on peanut and soil calcium. *Commun. Soil Sci. Plant Anal.* 20:1727–1744.

Alwan, A.H., H.A. Saadalla, and A.A. Muhammad. 1989. Preliminary study of the effect of salinity on germination, growth, and some mineral composition of different varieties of corn. *Ann. Agric. Sci.* (Cairo) 34:193–204.

Amann, C., and A. Amberger. 1989. Phosphorus efficiency of buckwheat, *Fagopyrum esculentum. Z. Pflanzeneraehr Bodenkd.* 152:181–189.

Amara, D.S., W.A. Mohjadi, and R.H. Miller. 1981. Tolerances of *Rhizobium phaseoli* to acidity, aluminum, and manganese. *Mircen. J.* 2:289–299.

Anderson, D.L., D.B. Jones, and G.H. Snyder. 1987. Response of rice-sugar cane rotation to calcium silicate slag on Everglade histosols. *Agron. J.* 79: 531–535.

Anderson, M. 1988. Toxicity and tolerance of aluminum in vascular plants. *Water, Air, Soil Pollution* 39:439–462.

Andrew, C.S. 1978. Mineral characterization of tropical forage legumes. *In* C.S. Andrew and E.J. Kamprath (Eds.). *Mineral Nutrition of Legumes in Tropical and Subtropical Soils*. CSIRO, East Melbourne, Australia, pp. 93–112.

Arines, J., A. Vilarino, and M.J. Sainz. 1989. Effect of different inocula of vesicular arbuscular mycorrhizal fungi on manganese content and concentrations in red clover, *Trifolium pratense* L., plants. *New Phytol.* 112:215–220.

Arnon, D.I., W.E. Fratzke, and C.M. Johnson. 1942. Hydrogen ion concentration in relation to absorption of inorganic nutrients by higher plants. *Plant Physiol.* 17:515–524.

Asghar, J., M. Ranizan, and S. Ahmad. 1986. Effect of presorting soils treatment on the yield of wheat under drought conditions. *J. Agric. Res. Pakistan* 24: 305–312.

Babich, H., and G. Stotsky. 1986. Manganese toxicity to fungi: Influence of pH. *Bull. Environ. Contam. Toxicol.* 27:474–480.

Baker, T.G. 1988. Measurement of aluminum toxicity in acidic soils. *Tech. Report. Series. Victoria Dept. Agric.* and Rural Affairs, Melbourne, Australia, No. 150, pp. 1–15.

Baligar, V.C., J.H. Elgin, Jr., and C.D. Foy. 1989. Variability in alfalfa for growth and mineral uptake-efficiency ratios under aluminum stress. *Agron. J.* 81:223–229.

Baligar, V.C., T.B. Kinraide, R.J. Wright, O.L. Bennett, and M.D. Smedley. 1987. Aluminum effects on growth and P, Ca, and Mg uptake-efficiency in red clover cultivars. X Intl. Plant Nutr. Colloq., Rockville, Maryland. *J. Plant Nutr.* 10:1131–1137.

Bataglia, O.C., C.E.O. Camargo, O.F. Oliveira, V. Nagiu, and V.J. Ramos. 1985. Response to liming of three wheat cultivars with different tolerance of aluminum toxicity. *Revista Brasileira De Ciencia Do Solo* 9:139–147.

Bell, L.C., and D.G. Edwards. 1986. The role of aluminum in acid soil infertility. *In* M. Latham (Ed.). *Soil Management Under Humid Conditions*. Intl. Board for Soil Research and Mgt., Inc., Bangkok, Thailand, Proc. No. 5, pp. 201–223.

Bell, R.W., D.G. Edwards, and C.J. Asher. 1989. External calcium requirements for growth and nodulation of six tropical food legumes grown in flowing nutrient culture. *Aust. J. Agric. Res.* 40:85–96.

Bengtsson, B., H. Asp, P. Jensen, and D. Berggren. 1988. Influence of aluminum on phosphate and calcium uptake in beech, *Fagus sylvatica* grown in nutrient solution and soil solution. *Physiol. Plant.* 74:299–305.

Berzonsky, W.A., and G. Kimber. 1989. The tolerance to aluminum of *Triticum* N-genome amphiploids. *Plant Breeding*. Paul Parey Publishers, Berlin and Hamburg, 103:37–42.

Bieka, S., and M. Kuchler. 1980. Investigations on aluminum tolerance in winter rye in a program of drought tolerance research. 2. Investigation on vernalized plants. *Archiv. Fur. Zuchtungforschung* 18:25–29.

Bilski, J.J. 1987. Response of plants to mineral stresses caused by acidification and salinization of the medium. I. Effect of the reaction of the medium and the aluminum and calcium contents of the growth and chemical composition of root systems of lupines and lucerne in water culture. *Burletyn Instytutu Hodowli Aklimatyzacji*, Roslin No. 164, pp. 199–206.

Bilski, J.J. 1988. Response of plants to mineral stress caused by acidification and salinizationm of the medium. II. The effect of aluminum and manganese on seedlings of some plants in a pot experiment. *Burletyn Instytutu Hodowli Aklimatyzacji*, Roslin No. 165, pp. 67–73.

Birch, H.F. 1953. Phosphate response, base saturation, and silica relationships in acid soils. *J. Agric. Sci.* 43:229–235.

Black, A.S., and K.C. Cameron. 1984. Effect of leaching on soil properties and lucerne growth following lime and gypsum amendments to a soil with an acid subsoil. *New Zeal. J. Agric. Res.* 27:195–200.

Blair, G.J., K.B. Lithgow, and P.W. Orchard. 1988. The effects of pH and calcium on the growth of *Leucaena leucocephala* in an oxisol and ultisol soil. *Plant Soil* 106:209–214.

Blamey, F.C.P., and K. Nathanson. 1977. Relationships between aluminum toxicity and sunflower yields on an Avalon, medium sandy loam. *Agrochemophysics* 9:59–66.

Blamey, F.P.C., D.G. Edwards, C.J. Asher, and M.K. Kun. 1982. Response of sunflower to low pH. *In* A. Scaife (Ed.). *Ninth Intl. Plant Nutr. Colloq.*, Warwick, England. Commonwealth Agric. Bureaux, Farnham House, Farnum Royal Slough, United Kingdom, Vol. 1, pp. 66–71.

Borgonovi, R.A., R.E. Schaffert, and G.V.E. Pitta. 1984. Breeding aluminum tolerent sorghums. *In Sorghums for Acid Soil.* Workshop Proc., Cali, Columbia, May 25-June 2, 1984, pub. jointly by INTSORMIL-ICRISAT & CIAT, tech. editors, L.M. Gourley and J.G. Salinas, pp. 271–292.

Boswell, F.C., K. Ohki, M.B. Parker, L.M. Shuman, and D.O. Wilson. 1981. Methods and rates of applied manganese for soybeans. *Agron. J.* 73:909–912.

Bouton, J.H., M.E. Sumner, J.E. Hammel, and H. Shahandeh. 1986. Yield of an alfalfa germplasm selected for acid soil tolerance when grown in soils with modified subsoils. *Crop Sci.* 26:334–336.

Bowen, J.E. 1969. Absorption of copper, zinc and manganese by sugarcane leaf tissue. *Plant Physiol.* 44:255–261.

Bromfield, E.S.P., and D.G. Jones. 1980. Studies on acid tolerance of *Rhizobium trifolii* in culture and soil. *J. Appl. Bact.* 48:253–264.

Brooke, H.D., D.R. Coventry, T.G. Reeves, and D.K. Jarvis. 1989. Liming and deep ripping responses for a range of field crops. *Plant Soil* 115:1–6.

Bruce, R.C., L.A. Warrell, D.G. Edwards, and L.C. Bell. 1988. Effects of aluminum and calcium in the soil solution of acid soil on root elongation of *Glycine max* Cv. Forrest. *Aust. J. Agric. Res.* 39:319–338.

Bryan, O.C. 1923a. Effect of reaction on growth, nodule information, and calcium content of alfalfa, alsike clover, and red clover. *Soil Sci.* 15:23–36.

Bryan, O.C. 1923b. Effect of acid soils on nodule forming bacteria. *Soil Sci.* 15:37–40.

Bueno, N., H.P. Haag, J.P. Pereire, I.J.M. Viegas, and N.E. Dende. 1985. Mineral nutrition of rubber. IX. Substrate aluminum affecting the development of toxic symptoms and concentrations in rubber (*Hevea* sp.) (Pt. Eng. sum.). *Anais Da Escola Superior Agricultura*, Luiz De Queiroz Universidade de Sao Paulo 45:319–339.

Burmester, C.H., J.F. Adams, and J.W. Odom. 1988. Response of soybeans to lime and molybdenum on ultisols of northern Alabama. *Soil Sci. Soc. Am. J.* 52:1391–1394.

Camargo, C.E.O., and J.C. Felicio. 1987. Trigo tricale centio availacao da efficiencia ao fosforo e tolerancia a toxidade ao aluminio. *Bragantia* (Campinas) 46:203–215.

Campbell, T.A., and C.D. Foy. 1987. Selection of grain amaranthus species for tolerance to excess aluminum in an acid soil. *J. Plant Nutr.* 10:249–260.

Campbell, T.A., N.J. Nuernberg, and C.D. Foy. 1989. Differential responses of alfalfa cultivars to aluminum stress. *J. Plant Nutr.* 12:291–305.

Caradus, J.R., and R.W. Snaydon. 1986. Plant factors influencing phosphorus uptake by white clover, *Trifolium repens*, from solution culture. II. Root and shoot primary and split root studies. *Plant Soil* 93:165–174.

Carvalho, M.M., D.G. Edwards, C.S. Andrew, and C.J. Asher. 1981. Aluminum toxicity, nodulation, and growth of *Stylosanthes* species. *Agron. J.* 73:261–265.

Cassel, D.K. 1980. Effects of plowing depth and deep incorporation of lime and phosphorus upon physical and chemical properties of two coastal plain soils after 15 years. *Soil Sci. Soc. Am. J.* 44:89–95.

Cheng, B.T. 1982. Some significant functions of silicon in higher plants. *J. Plant Nutr.* 5:1345–1353.

Chong, K., J.C. Wynne, G.H. Elkins, and T.J. Schneeweis. 1987. Effects of soil acidity and aluminum content on *Rhizobium* inoculation, growth, and nitrogen fixation of peanuts amd other grain legumes. *Trop. Agric.* (Trinidad) 64:97–104.

Christiansen, M.N., H.R. Carns, and O.J. Slyter. 1970. Stimulation of solute loss from radicles of *Gossypium hirsutum* L. by chilling, anaerobiosis, and low pH. *Plant Physiol.* 46:53–56.

Clark, R.B. 1982. Plant response to mineral element toxicity and deficiency. *In* M.N. Christensen and C.F. Lewis (Eds.). *Breeding Plants for Less Favorable Environments*. John Wiley and Sons, New York, pp. 71–142.

Clark, R.B. 1984. Physiological aspects of calcium, magnesium, and molybdenum deficiencies in plants. *In* F. Adams (Ed.). *Soil Acidity and Liming*, Second Edition. Soil Sci. Soc. Am., Amer. Soc. Agron. and Crop Sci. Soc. Am., Madison, Wisconsin, pp. 99–170.

Clark, R.B., and L.M. Gourley. 1988. Mineral element concentration of sorghum senotypes grown on tropical acid soil. *Commun. Soil Sci. Plant Anal.* 19:1019–1029.

Clark, R.B., C.I. Flores, and L.M. Gourley. 1988. Mineral element concentrations in acid soil tolerant and susceptible sorghum genotypes. *Commun. Soil Sci. Plant Anal.* 19:1003–1017.

Clark, R.B., H.A. Pier, D. Knudsen, and J.W. Maranville. 1981. Effect of trace element deficiencies and excesses on mineral nutrients in sorghum. *J. Plant Nutr.* 3:357–374.

Classens, A.S., and W.J. Folscher. 1988. The influence of different phosphorus and amonium and nitrate supply levels on the yield and phosphorus and calcium relationships in wheat. *S. Afr. J. Plant Soil* 5:127–130.

Coltman, R.R. 1987. Tolerance of tomato strains to phosphorus deficiency in nutrient culture. Hort. Sci. 22:1305–1307.

Cooper, J.E., M. Wood, and A.J. Bjourson. 1986. Influence of acidity and aluminum on the nodulation of marsh trefoil (*Lotus pedunculatus*) and white clover (*Trifolium repens*) by rhizobia. Hort. Sci. 21:262–263.

Coventry, D.R., G.R. Morrison, T.G. Reeves, J.R. Hirth, and K.K.H. Fung 1987. Mineral composition and response to fertilizers of wheat grown on limed and deep ripped soil in northeastern Victoria, Australia. *Aust. J. Exp. Agric.* 27:687–694.

Cox, D.A. 1988. Lime, molybdenum, and cultivar effects on molybdenum deficiencies of poinsettia. *J. Plant Nutr.* 11:589–603.

Cregan, P.D. 1980. Soil acidity and associated problems. Guidelines for farmer recommendations. *AG Bull.* Wagga Wagga Dept. of Agric., New South Wales, Australia, October, 7, 1980.

Cumming, J.R., and L.H. Weinstein. 1990. Aluminum—mycorhizal interactions in the physiology of pitch pine seedlings. *Plant Soil* 125:7–18.

DeMiranda, L.N., and D.L. Rowell. 1987. The effects of lime and phosphorus on the function of wheat roots in acid top soils and subsoils. *Plant Soil* 104: 253–262.

Devine, T.E., C.D. Foy, A.L. Fleming, C.H. Hanson, T.A. Campbell, J.E. McMurtrry, III, and J.W. Schwartz. 1976. Development of alfalfa strains with different tolerance to aluminum toxicity. *Plant Soil* 44:73–79.

Dobereiner, J. 1966. Manganese toxicity effects on nodulation and nitrogen fixation of beans. *Phaseolul vulgaris* L.) in acidic soils. *Plant Soil* 24:153–166.

Doolas, G.G. 1930. Soil acidity and soybean inoculation. *Soil Sci.* 30:275–288.

Doss, B.D., and Z.F. Lund. 1975. Subsoil pH effects on growth and yield of cotton. *Agron. J.* 67:193–196.

Doss, B.D., W.T. Dumas, and Z.F. Lund. 1979. Depth of lime incorporation for correction of subsoil acidity. *Agron. J.* 71:541–544.

Duncan, R.R. 1981. Registration of GPIR acid soil tolerant sorghum germplasm population. (Reg. No. GP73). *Crop Sci.* 21:637.

Duncan, R.R. 1984. Registration of acid soil tolerant sorghum germplasm (Reg. Nos. GR 140–142) *Crop Sci.* 24:1006.

Duncan, R.R. 1988. Sequential development of acid soil tolerant genotypes under field stress conditions. *Commun. Soil Sci. Plant Anal.* 19:1295–1305.

Duncan, R.R., S. Torres-Cardona, R. Goenaga, A. Sotomayor-Rios, and L.M. Shuman. 1989. Registration of GPP4BR (H)C5 acid soil tolerant sorghum germplasm population. *Crop Sci.* 29:1581–1582.

Ebelhar, M.W., and G.R. Tupper. 1985. Subsoil lime placement for cotton production on problem acid soils. *Proc. Beltwide Cotton Prod. Res. Conf.*, National Cotton Council Am., pp. 300–302.

Edwards, J.H., B.D. Horton, A.W. White, Jr., and O.L. Bennett. 1985a. Fluidized bed combustion residue as an alternative liming material and calcium source. *Commun. Soil Sci. Plant Anal.* 16:621–637.

Edwards, J.H., A.W. White, Jr., and O.L. Bennett. 1985b. Effects of fluidized bed combustion residue on pecan seedling growth and nutrient content. *Commun. Soil Sci. Plant Anal.* 16:639–652.

Ekdahl, I. 1957. The growth of root hairs and roots in nutrient media and distilled water and the effects of oxylate. *Lantbrukshoegsk Ann.* 23:497–518.

Elgawad, S.H., G.H. Gascho, and J.J. Sheet. 1982a. Response of sugar cane to silicate source and rate. I. Growth and yield. *Agron. J.* 74:481–484.

Elgawad, S.H., G.J. Gascho, and J.J. Sheet. 1982b. Response of sugar cane to silicate source and rate. II. Leaf freckling and nutrient content. *Agron. J.* 74:484–487.

Evans, C.E., and E.J. Kamprath. 1970. Lime response as related to percent Al saturation, solution aluminum, and organic matter content. *Soil Sci. Soc. Am. Proc.* 34:893–896.

Fageria, N.K. 1985. Influence of aluminum in nutrient solutions on chemical composition in two rice cultivars at different growth stages. *Plant Soil* 85: 423–429.

Fageria, N.K., and V.C. Baligar. 1989. Response of legumes and cereals to phosphorus in solution culture. *J. Plant Nutr.* 12:1005–1019.

Fageria, N.K., and M.P.B. Filho. 1987. Phosphorus fixation in oxisol of Brazil. *Fert. Agric. O* (94):33–38. Nat. Rice Bean Res. Centerm, EMBRAPA, Brazil.

Fageria, N.K., V.C. Baligar, and R.J. Wright. 1989. Growth and nutrient concentrations of alfalfa and common bean as influenced by soil acidity. *Plant Soil* 119:331–335.

Fageria, N.K., R.J. Wright, and V.C. Baligar. 1988a. Rice culture response to aluminum in nutrient solutions. *Commun. Soil Sci. Plant Anal.* 19:1133–1142.

Fageria, N.K., O.P. Morais, V.C. Baligar, and R.J. Wright. 1988b. Response of rice cultivars to phosphorus supply in an oxisol. *Fert. Res.* 16:195–206.

Farina, M.P.W., and P. Channon. 1980. Acid subsoil amelioration. I. A comparison of several mechanical procedures. *Soil Sci. Soc. Am. J.* 52: 169–175.

Farina, M.P.W., and P. Channon. 1988. Acid subsoil amelioration. II. Gypsum effects on growth and subsoil chemical properties. *Soil Sci. Soc. Am. J.* 52: 175–180.

Farin, M.P.W., P. Mendes, H.O. Gevers, and P. Channon. 1982. Differential tolerance to soil acidity among several South African maize genotypes. *Crop Prod.* 11:133–139.

Fernandez, L.R., and W.G. Blue. 1984. Reduction of subsoil acidity by surface applied amendments on an entisol and an ultisol from Florida. *Soil and Crop Sci. Soc. Fla. Proc.* 43:26–34.

Ferreiro, E.A., A.K. Helmy, and S.G. De Bussetti. 1985. Molybdate sorption by oxides of aluminum and iron. *Z. Pflanzenernaehr Bodenk.* 148:559–566.

Fey, M.V., A.D. Noble, and M.E. Sumner. 1989. Reply to comments on the calcium aluminum balance (CAB). *Soil Sci. Soc. Am. J.* 53:1920.

Fist, A.J., F.W. Smith, and D.G. Edwards. 1987. External phosphorus requirements of five tropical grain legumes grown in flowing solution culture. *Plant Soil* 99:75–84.

Fleming, A.L. 1989. Enhanced Mn accumulation by snap bean cultivars under Fe stress. *J. Plant Nutr.* 12:715–731.

Floyd, M., and S. Mookherji. 1987. Nitrogen fixation. *Soybean Newsletter* 14:234–236.

Foehse, D., N. Claasen, and A. Jungk. 1988. Phosphorus efficiency of plants. I. External and internal phosphorus requirements and phosphorus uptake efficiencies of different plant species. *Plant Soil* 110:101–110.

Fox, R.H., and W.P. Piekielek. 1984. Soil magnesium level, corn yield and magnesium uptake. *Commun. Soil Sci. Plant Anal.* 15:109–123.

Fox, R.L. 1979. Comparative responses of field grown crops to phosphate concentrations in soil solutions. *In* H. Mussell and R.C. Staples (Eds.). *Stress Physiology in Crop Plants*. John Wiley and Sons, Inc., New York, pp. 81–106.

Fox, R.L. 1981. External phosphorus requirements of crops. *Biochem. Soil Environ.*, ASA Special Pub. (AASPC3), Madison, Wisconsin, pp. 223–239.

Fox, R.L., W.M.H. Saunders, and S.S.S. Rojam. 1986. Phosphorus nutrition of pasture species: Phosphorus requirement and root saturation values. *Soil Sci. Soc. Am. J.* 50:142–147.

Fox, R.L., J.A. Silva, O.R. Younger, D.L. Plucknett, and G.D. Sherman. 1967. Soil and plant silicon and silicate response by sugar cane. *Soil Sci. Soc. Am. Proc.* 31:775–779.

Foy, C.D. 1964. Toxic factors in acid soils of the southeastern United States as related to the response of alfalfa to lime. *USDA Prod. Res. Report 80*, U.S. Govt. Printing Office, Washington, D.C.

Foy, C.D. 1973. Manganese and plants. *In Manganese*. Natl. Acad. Sci., Natl. Res. Council, Washington, D.C., pp. 51–76.

Foy, C.D. 1974a. Effects of aluminum on plant growth. *In* E.W. Carson (Ed.). *The Plant Root and Its Environment*. University Press of Virginia, Charlottesville, pp. 601–642.

Foy, C.D. 1974b. Effects of soil calcium availability on plant growth. *In* E.W. Carson (Ed.). *The Plant Root and Its Environment*, Un. Press of VA, Charlottesville, VA, pp. 565–600.

Foy, C.D. 1983. Plant adaptation to mineral stress in problem soils. *Iowa J. Res.* 57:339–354.

Foy, C.D. 1984. Physiological effects of hydrogen, aluminum, and manganese toxicities in acid soils. *In* F. Adams (Ed.). *Soil Acidity and Liming*, Second

Edition, Soil Sci. Soc. Am., Amer. Soc. Agron., and Crop Sci. Soc. Am., Madison, Wisconsin, pp. 57–97.

Foy, C.D. 1987. Acid soil tolerances of two wheat cultivars related to soil pH, KCl extractable aluminum, and aluminum saturation. *J. Plant Nutr.* 10: 609–623.

Foy, C.D. 1988. Plant adaptation to acid, aluminum toxic soils. *Commun. Soil Sci. Plant Anal.* 19:959–987.

Foy, C.D., and T.A. Campbell. 1984. Differential tolerances of amaranthus strains to high levels of aluminum and manganese in acid soils. *J. Plant Nutr.* 7:1365–1388.

Foy, C.D., and A.L. Fleming. 1982. Aluminum tolerance of two wheat cultivars related to nitrate reductase activities. *J. Plant Nutr.* 5:1313–1333.

Foy, C.D., and A.J. Oakes. 1984. A winter hardy, aluminum tolerant perennial pasture grass for reclamation of acid mine spoils. *J. Plant Nutr.* 7:929–951.

Foy, C.D., C.L. Dewald, and W.A. Berg. 1989. Differential tolerances of two Old World bluestem genotypes to excess aluminum in acid soil and nutrient solution. *J. Plant Nutr.* 12:1231–1254.

Foy, C.D., J.B. Scott, and J.A. Fisher. 1988. Genetics and breeding of plants tolerant of manganese toxicity. *In* R.D. Graham, R.J. Hannam, and N.C. Uren (Eds.). *Manganese in Soils and Plants*. Kluwer Academic Publishers, Boston, pp. 293–307.

Franco, A.A., and J. Dobereiner. 1971. Manganese toxicity in an acid soil affecting the soybean *Rhizobium* symbiosis. *Pesq. Agropec. bras. ser. Agron.* 6:57–66.

Franco, A.A., and D.N. Munns. 1982. Acidity and aluminum restraints on nodulation, nitrogen fixation, and growth of *Phaseolus vulgaris* in nutrient solution. *Soil Sci. Soc. Am. J.* 46:296–301.

Fuller, R.D., and C.J. Richardson. 1986. Aluminate toxicity as a factor controlling plant growth in bauxite residue. *Environ. Toxicol. Chem.* 5:905–915.

Furlani, R.R., and R.B. Clark. 1981. Screening sorghum for aluminum tolerance in nutrient solution. *Agron. J.* 73:587–594.

Gallo, P.B., H.A.A. Mascarenhas, J.A. Quaggio, and O.C. Bataglia. 1986. Differential responses of soybeans and sorghum to liming. *Rev. Bras. De Ciencia Do Solo* 10:253–258.

Galvez, L., R.B. Clark, L.M. Gourley, and J.M. Maranville. 1989. Effects of silicon and mineral composition of sorghum grown with excess manganese. *J. Plant Nutr.* 12:547–561.

Gillman, G.P., and M.E. Sumner. 1987. Surface charge characterization and soil solution composition of four soils from the Southern Piedmont of Georgia. *Soil Sci. Soc. Am. J.* 51:589–594.

Goldman, I.L., T.E. Carter, Jr., and R.P. Patterson. 1989a. Differential genotypic response to drought stress and subsoil aluminum in soybean. *Agron. J.* 81:330–334.

Goldman, I.L., T.E. Carter, Jr., and R.P. Patterson. 1989b. A detrimental interaction of subsoil aluminum and drought stress on the leaf water status of soybean. *Agron. J.* 81:461–463.

Gotbold, D.L., K. Dictus, and A. Huettermann. 1988. Influence of aluminum and nitrate on root growth and mineral nutrition of Norway spruce *Picea abies* seedlings. *Can. J. For. Res.* 18:1167–1171.

Gourley, L.M. 1984. Finding and utilizing exotic Al-tolerant sorghum germplasm. *In* L.M. Gourley and J.G. Salinas (Eds.). *Sorghums For Acid Soils*. Workshop Proc., INTSORMIL, ICRISAT, CIAT, Cali, Colombia, May 28–June 2, 1984, pp. 293–309.

Gourley, L.M. 1987. Identifying aluminum tolerance in sorghum genotypes grown on tropical acid soils. *In* W.H. Gabelman and B.C. Loughman (Eds.). *Genetic Aspects of Mineral Nutrition*. Martinus Nijhoff, Boston, pp. 89–98.

Graham, P.H., and C.A. Parker. 1964. Diagnostic features in the characterization of root nodule bacteria of legumes. *Plant Soil* 20:383–396.

Graham, R.H., R.J. Hannam, and N.C. Uren. (Eds.). 1988. *Manganese In Soils and Plants*. Kluwer Academic Publishers, Boston, Dordrecht, London.

Grimme, H. 1984. Aluminum tolerance of soybean plants as related to magnesium nutrition. *Proc. VI Intl. Colloq. for Optimization of Plant Nutrition*, Vol. 1. Montpelier Cedex, France, pp. 243–249.

Habben, J.E., and D.G. Blevins. 1989. Transfer of Rb-86 (potassium) from deeply rooted alfalfa (*Medicago sativa* L.) to associated shallow rooted maize (*Zea mays* L.) or grain sorghum (*Sorghum vulgare* Pers.). *J. Plant Nutr.* 12:1089–1104.

Hagedorn, C., and B.A. Caldwell. 1981. Characterization of diverse *Rhizobium trifolii* isolates. *Soil Sci. Soc. Am. J.* 45:513–516.

Hai, T.V., T.T. Nga, and H. Laudelot. 1989. Effect of aluminum on the mineral nutrition of rice. *Plant Soil* 114:173–185.

Hallock, D.L. 1979. Relative effectiveness of several Mn sources on Virginia type peanuts. *Agron. J.* 71:685–688.

Hanson, J.B. 1984. The functions of calcium in plant nutrition. *In* P.B. Tinker and A. Lauchli (Eds.). *Advances in Plant Nutrition*, Vol. 1. pp. 149–208.

Hartel, P.G., and J.H. Bouton. 1989. *Rhizobium meliloti* inoculation of alfalfa selected for tolerance to acid, aluminum rich soils. *Plant Soil* 116:283–285.

Helal, H.M., and A. Dressler. 1989. Mobilization and turnover of soil phoshorus in the rhizosphere. *Z. Pflanzenernaehr Bodenkd.* 152:175–180.

Hern, J.L., H.A. Menser, R.C. Sidle, and T.E. Staley. 1982. Effects of surface applied lime and EDTA on subsoil acidity and aluminum. *Soil Sci.* 145:52–57.

Hill, P.R., J.L. Aldrichs, and G. Ejeta. 1989. Rapid evaluation of sorghum for aluminum tolerance. *Plant Soil* 114:85–90.

Horiguchi, T. 1987. Mechanism of manganese toxicity and tolerance in plants. II. Deposition of oxidized manganese in plant tissues. *Soil Sci. Plant Nutr.* 33: 595–600.

Horiguchi, T. 1988. Mechanism of manganese toxicity and tolerance in plants. IV. Effects of silicon and alleviation of manganese toxicity of rice plants. *Soil Sci. Plant Nutr.* 34:65–73.

Howard, D.D., and F. Adams. 1965. Calcium requirement for penetration of subsoils by primary cotton roots. *Soil Sci. Soc. Am. Proc.* 29:558–562.

Hoyt, P.B., and G.G.S. Myovella. 1979. Correction of severe manganese deficiency in wheat with chemical fertilizers. *Plant Soil* 52:437–444.

Hue, N.V. 1988. A possible mechanism for manganese phytotoxicity in Hawaii, USA soils amended with a low manganese sewage sludge. *J. Environ. Qual.* 17:473–479.

Hue, N.V., G.R. Craddock, and F. Adams. 1986. Effect of organic acids on aluminum toxicity in subsoils. *Soil Sci. Soc. Am. J.* 50:28–34.

Hussain, F., R. Overstreet, and L. Jacobson. 1954. The influence of hydrogen ion concentration on cation absorption by barley roots. *Plant Physiol.* 29:234–237.

Hutton, E.M. 1985. Centrosema breeding for acid tropical soils with emphasis on efficient Ca absorption. Trop. *Agric.* (Trinidad) 62:273–280.

Inoue, K., H. Yohota, and Y. Yamada. 1988. Effect of calcium in the medium on root growth under low pH conditions. *Soil Sci. Plant Nutr.* 34:359–374.

IRRI, 1974. *International Rice Research Institute Report for 1973*. IRRI, Los Banos, Philippines.

Islam, A.K.M.S., D.G. Edwards, and C.J. Asher. 1980. pH optima for crop growth: Results of a flowing nutrient culture experiment with six species. *Plant Soil* 54:339–357.

Jadi, A.W.M. 1982. Some abiotic factors affecting the survival of *Rhizobium* in soil. *In* Z. Sham Suddin (Ed.). *Proc. Sem. Nitrogen in Malaysian Agric.* Kuala Lampur, Malaysia, pp. 29–42.

Jan, F., and S. Pettersson. 1989. Varietal diversity of upland rice in sensitivity to aluminum. *J. Plant Nutr.* 12:973–993.

Jarrell, W.M. 1989. Effect of saline irrigation waters on soil manganese leaching and bioavailability to sugar beet. *Soil Sci. Soc. Am. J.* 53:142–146.

Jarvis, S.C. 1984. The response of two genotypes of white clover to addition of lime to an acid, permanent grassland soil. *J. Sci. Food Agric.* 35:1149–1158.

Jarvis, S.C., and L.H.P. Jones. 1987. The absorption and transport of manganese by perennial ryegrass and white clover as affected by silicon. *Plant Soil* 99: 231–240.

Jones, D., M.H.R. Browning, and T.C. Hutchinson. 1986. The influence of mycorrhizal association on paper birch and jack pine seedlings when exposed to elevated copper, nickel or aluminum. *Water Air Soil Pollut.* 31:411–448.

Jones, L.H. 1961. Aluminum uptake and toxicity in plants. *Plant Soil* 13:292–310.

Jones, R.S., and C.A. Mitchell. 1989. Calcium involvement in growth inhibition of mechanically stressed soybean *Glycine max* seedlings. *Plant Physiol.* 76: 598–602.

Kamprath, E.J., and C.D. Foy. 1985. Lime-fertilizer-plant interactions in acid soils. *In* O.P. Engelstad (Ed.). *Fertilizer Technology and Use*, Third Edition. Soil Sci. Soc. Am., Madison, Wisconsin, pp. 91–151.

Kapoor, M.L., G.S. Seklion, and B. Sirrgh. 1982. Nitrogen, phosphorus, and postassium nutrition of maize and wheat from two surface subsurface soil layers. *Ind. J. Biol.* 9:264–269.

Karanja, N.K., and M. Wood. 1988. Selecting *Rhizobium phaseoli* strains for use with beans, *Phaselous vulgaris*, in Kenya: Infectiveness and tolerance of acidity and aluminum. *Plant Soil* 112:7–14.

Katsuhara, M., and M. Tazawa. 1987. ATP is essential for calcium induced salt tolerance in *Nitellopsis oblisa*. *Protoplasma* 138:190–192.

Kauffman, M.D., and E.H. Gardner. 1978. Segmental liming of soil and its effect on the growth of wheat. *Agron. J.* 70:331–336.

Kayda, M., and L. Zvacek. 1989. Aluminum and manganese and their relation to calcium in soil solutions and needles of three Norway spruce (*Picea abies* L. Karst) stands of upper Austria. *Plant Soil* 114:257–267.

Kennedy, C.W., W.C. Smith, Jr., and M.T. Ba. 1986. Root cation exchange capacity of cotton cultivars in relation to aluminum toxicity. *J. Plant Nutr.* 9:1123–1133.

Kennedy, C.W., M.T. Ba, A.G. Caldwell, R.L. Hutchinson, and J.E. Jones. 1987. Differences in root and shoot growth and soil moisture extraction between cotton cultivars in an acid subsoil. *Plant Soil* 101:241–246.

Kerridge, P.C. 1969. *Aluminum toxicity in wheat* (*Triticum aestivum* Vill. Host). Ph.D. Dissertation Oregon State University, Univ. Microfilms 69-4086, Ann Arbor, Michigan (Diss. Abstr. 293:3159).

Keyser, H.J., and D.N. Munns. 1979. Effects of calcium, manganese, and aluminum on growth of rhizobia in acid media. *Soil Sci. Soc. Am. J.* 43: 500–503.

Keyser, H.J., D.N. Munns, and J.S. Hohenberg. 1979. Acid tolerance of rhizobia in culture and in symbiosis with cowpea. *Soil Sci. Soc. Am. J.* 43:719–722.

Kinraide, T.B., and D.R. Parker. 1987. Non-phytotoxicity of the aluminum sulfate ion. *Plant Physiol.* 71:207–212.

Kinraide, T.B., and D.R. Parker. 1989. Comments on the calcium aluminum balance (CAB). *Soil Sci. Soc. Am. J.* 53:1919–1920.

Kirkby, E.A., and D.J. Pilbeam. 1984. Calcium as a plant nutrient. *Plant Cell Environ.* 7:207–212.

Klimov, S.V. 1985. Interactions of stress factors: Intensification of drought action on plants when Al^{3+} is present in the medium. *Sov. Plant Physiol.* 32:423–428.

Klotz, F., and W.J. Horst. 1988. Genotypic differences in aluminum tolerance of soybean (*Glycine max* L.) as affected by ammonium and nitrate nitrogen nutrition. *Plant Physiol.* 132:702–707.

Korner, J.E., I.M. Moller, and P. Jensen. 1987. Effects of calcium and other divalent cations on uptake of nickel by excised barley roots. *Plant Physiol.* 71:49–54.

Koslowsky, S.D., and R.E.J. Boermer. 1989. Integrated effects of aluminum, phosphorus, and mycorrhizae on growth and nutrient uptake of *Panicum virgatum*. *Environ. Poll.* 61:107–126.

Kpoghomon, B., V.T. Sapra, B.P. Singh, M. Rangappa, M.E. Kraemer, A. Bhagsan, M. Reddy, R.P. Pacumbaba, M. Floyd, S.C. Tewari, and P.W. Igbokwe. 1986. Evaluation of soybean germplasm for stress tolerance and biological efficiency. *Soybean Genetics Newsletter* 13:186–197.

Krizek, D.T., and C.D. Foy. 1988a. Role of water stress in differential aluminum tolerance of two barley genotypes grown in acid soil. *J. Plant Nutr.* 11:351–367.

Krizek, D.T., and C.D. Foy. 1988b. Mineral element concentrations of six sunflower cultivars in relation to water deficit and aluminum toxicity. *J. Plant Nutr.* 11:409–422.

Krizek, D.T., and C.D. Foy. 1988c. Mineral element concentrations of two barley cultivars in relation to water deficit and alumimim toxicity. *J. Plant Nutr.* 11:369–386.

Krizek, D.T., C.D. Foy, and W.P. Wergin. 1988. Role of water stress in differential aluminum tolerance of six sunflower cultivars grown in acid soil. *J. Plant Nutr.* 11:387–408.

Kroetz, M.E., W.H. Schmidt, J.E. Beurlein, and G.L. Ryder. 1977. Controlling manganese deficiency increases soybean yields. *Ohio Report* 62:51–53.

Kruger, E., and E. Sucoff. 1989. Growth and nutrient status of *Quercus rubra* L. in response to Al and Ca. *J. Exp. Bot.* 40:653–658.

Kucey, R.M.N., H.H. Janzen, and M.E. Leggett. 1989. Microbially mediated increases in plant available P. *Adv. Agron.* 42:199–228.

Kummerova, M., and I. Buresova. 1989. Manganese effect on the growth of root and hypocotyl of *Latuca sativa* L. *Ser. Fac. Sci. Nat. Univ. Purkynianae Brun* 19:55–62.

Kundu, S., M.B. Kamath, and N.N. Goswami. 1986. Effect of silicate on phosphate utilization by wheat. *J. Nucl. Agric. Biol.* 15:108–114.

Kurashima, K., and T. Kimura. 1987. The influence of exchangeable soil aluminum on magnesium uptake by cocksfoot. *Bull. Nat. Grassl. Res. Inst.* 36:28–34.

Lafever, H.N. 1988. Registration of Cardinal wheat. *Crop Sci.* 28:377.

Li, Y.C., A.K. Alva, and M.E. Sumner. 1989. Response of cotton cultivars to aluminum in solutions with varying silicon concentrations. *J. Plant Nutr.* 12:881–892.

Lindstrom, K., and H. Myllyniemi. 1987. Sensitivity of red clover rhizobia to soil acidity factors in pure culture and in symbiosis. *Plant Soil* 98:353–362.

Little, R. 1988. Plant soil interactions at low pH: Problem solving—the genetic approach. *Commun. Soil Sci. Plant Anal.* 19:1239–1257.

Long, F.L., and C.D. Foy. 1970. Plant varieties as indicators of aluminum toxicity in the A2 horizon of a Norfolk soil. *Agron. J.* 62:679–681.

Long, O.H. 1959. Root studies on some farm crops in Tennessee. *University of Tennessee Agr. Exp. Stn. Bull. 301*, 1–41, Knoxville.

Lu, N., and J.H. Edwards. 1985. Injection of chemical amendments into compacted subsoils. l. Sorghum, *Sorghum bicolor. Commun. Soil Sci. Plant Anal.* 16:1015–1028.

Lu, N., and J.H. Edwards. 1986. Injection of chemical amendments into compacted subsoils, growth, and nutrient uptake by wheat, *Triticum aestivum. Commun. Soil Sci. Plant Anal.* 17:1055–1070.

Lund, Z.F. 1970. The effect of calcium and its relation to several cations in soybean root growth. *Soil Sci. Soc. Am. Proc.* 34:456–459.

Lynch, L., G.R. Cramer, and A. Lauchli. 1987. Salinity reduces membrane associated calcium in corn root protoplasm. *Plant Physiol.* 83:390–394.

Maas, E.V. 1969. Calcium uptake by excised maize roots and interactions with alkali cations. *Plant Physiol.* 44:985–989.

Maas, E.V., and C.M. Grieve. 1987. Sodium induced calcium deficiency in salt stressed corn. *Plant Cell Environ.* 10:559–564.

Mallett, J.B., B. St. E. Clemence, M.P.W. Farina, and S.M. Beghui. 1987. A theoretical determination of the optimum depth of lime incorporation in soil with subsoil aluminum toxicity problems. *Applied Science.* Sunnyside South African Weed Soc., Vol. 1 (1):23–27.

Manson, A.D., and M.V. Fey. 1989. Cation type and ionic strength effects in the solution composition of an acidic subsoil. *J. Soil Sci.* 40:577–584.

Manrique, L.A. 1986. Effect of extreme soil acidity on plant growth and yield of cassava. *Trop. Root Tuber Crops Newsletter* No. 16–17:27–36.

Marks, M.J., and G.C. Soane. 1987. Crop and soil response to subsoil loosening, deep incorporation of phosporus and potassium fertilizer, and subsequent soil management on a range of soil types. Part 1. Response of arable crops. *Soil Use Mgt.* 3:115–123.

Marschner, H. 1986. *Mineral Nutrition of Higher Plants.* Academic Press, London.

Martinez, M.V., and A. Cerda. 1989. Nitrate reductase activity in tomato and cucumber leaves as influenced by NaCl and N sources. *J. Plant Nutr.* 12: 1335–1350.

Mascarenhas, H.A.A., M.A.C. De Miranda, and E.A. Bulisani. 1985. Soybean tolerance to manganese in the tropics. *In* S. Shanmugasundaram and E.W. Surzberger (Eds.). *Soybean in Tropical and Subtropical Cropping Systems.* Shanna Taiwan Asian Veg. Res. Dev. Center, pp. 313–318.

Mayland, H.F., and D.L. Grunes. 1979. Soil-climate-plant relationships in the etiology of grass tetany. *In* V.V. Rendig and D.L. Grunes (Eds.). *Grass Tetany.* Amer. Soc. Agron., Madison, Wisconsin, pp. 123–175.

McCormick, L.H., and F.A. Amendale. 1983. Soil pH, extractable aluminum, and tree growth on mine soils. *Commun. Soil Sci. Plant Anal.* 14:249–262.

McEwen, J., and A. Johnson. 1979. The effects of subsoiling and deep incorporation of phosphorus and potassium fertilizers on the yield and mineral uptake of barley, potatoes, wheat, and sugar beet grown in solution. *J. Agron. Sci.* 92:695–702.

McKenzie, R.C., D.C. Penny, L.W. Hodgins, B.S. Anlakh, and H. Ukrainetz. 1988. The effects of liming on an ultisol in northern Gambia. *Commun. Soil Sci. Plant Anal.* 19:1355–1369.

McLaughlin, M.J., and T.R. James. 1989. The effect of subsurface aluminum on the growth and uptake of surface applied phosphorus by wheat seedlings. *J. Plant Nutr.* 12:603–620.

McLean, E.E., and M.D. Carbonell. 1972. Calcium, magnesium, and potassium saturation ratios in two soils and their effects upon yield and nutrient contents of German millet and alfalfa. *Soil Sci. Soc. Am. Proc.* 36:927–930.

McNaughton, S.J., J.L. Tarrants, M.M. McNaughton, and R.H. Davis. 1985. Silica as a defense against herbivores and a growth promoter of African grasses. *Ecology* 66:528–535.

Menge, J.A., R.M. Davis, E.L.V. Johnson, and G.A. Zentmeyer. 1978. Mycorrhizal fungi increase growth and reduce transplant injury in avocado. *Calif. Agric.* 32:6–9.

Meredith, C.P., A.J. Connor, and T.M. Schettini. 1988. The use of cell selection to obtain novel plant genotypes to mineral stress. *Iowa State J. Res.* 62: 523–535.

Messick, D.L. 1982. The movement of calcium and magnesium in related Virginia soil from surface applied dolomitic limestone. M.S. Thesis. Virginia Polytechnic Institute and State University, Blacksburg.

Miyake, Y., and E. Takahashi. 1985. Effect of silicon on the growth of soybean *Glycine max* plants in a solution culture. *Soil Sci. Plant Nutr.* 31:625–636.

Moore, D.P. 1974. Physiological effects of pH on roots. *In* E.W. Carson (Ed.). *The Plant Root and Its Environment.* University Press of Virginia, Charlottesville, pp. 135–151.

Moore, D.P., R. Overstreet, and L. Jacobson. 1961. Uptake of magnesium and its interaction with calcium in excised barley roots. *Plant Physiol.* 36:290–295.

Moore, P.A. Jr., and W.H. Patrick, Jr. 1989. Manganese availability and uptake of rice in acid sulfate soils. *Soil Sci. Soc. Am. J.* 53:104–109.

Moraghan, J.T. 1979. Manganese toxicity in flax grown on certain calcareous soils low in available iron. *Soil Sci. Soc. Am. J.* 43:1177–1180.

Moschler, W.W., G.D. Jones, and G.W. Thomas. 1960. Lime and soil acidity effects on alfalfa growth in red-yellow podzolic soil. *Soil Sci. Soc. Am. Proc.* 24:507–509.

Muchovej, R.M.C., V.G. Allen, D.C. Martens, and J.J. Muchovej. 1988. Effects of aluminum chelate in nutrient solution on the growth and composition of ryegrass. *J. Plant Nutr.* 11:117–129.

Muchovej, R.M.C., J.J. Muchovej, and V.H. Alvarez. 1989. Temporal relations of phosphorus fractions in an oxisol amended with rock phosphate and *Thiobacillus thioxidans*. Soil Sci. Soc. Am. J. 53:1096–1100.

Munns, D.N. 1965. Soil acidity and growth of a legume. 1. Interactions of lime with nitrogen and phosphate on growth of *Medicago sativa* L. and *Trifoliium subterraneum* L. *Aust. J. Agric. Res.* 16:733–741.

Munns, D.N. 1978. Legume-*Rhizobium* relations. *In* C.S. Andrew and E.J. Kamprath (Eds.). *Mineral Nutrition of Legumes in Tropical and Subtropical Soils*. CSIRO, E. Melbourne, Australia. pp. 247–263.

Munns, D.N. 1979. Tolerances of soil acidity in symbiosis of *Vignia radiata* with rhizobia. *Agron. J.* 71:256–260.

Muzilli, O. 1984. Soil management as an alternative for minimizing environmental constraints for wheat production in the semi-tropical areas of Brazil. *In Wheats For More Tropical Environments*. Proc. Intl. Symp., CIMMYT, Mexico, D. F., pp. 231–237.

Neves, J.C.L., R.F. Novais, and N.F. DeBarros. 1982a. Effect of aluminum in nutrient solution on growth and nutrient uptake of eucalyptus species. *Revista Arvore* 6:1–16. New Record. (Pt., Eng.).

Neves, J.C.L., N.F. DeBarros, R.F. Novais, J.L. DeAnjos, and J.L. Dos. 1982b. Effect of aluminum in samples of two Cerrado latosols on the growth and nutrient uptake of eucalyptus seedlings. *Revista Arvore* 6:17–28.

Njohn, B.O., W.O. Enwezor, and B.I. Onzenakwe. 1987. Calcium deficiency identified as an important factor limiting maize growth in acid ultisols of eastern Nigeria. *Fert. Res.* 14:113–124.

Noble, A.D., and M.E. Sumner. 1988. Calcium and aluminum interactions and soybean growth in nutrient solutions. *Commun. Soil Sci. Plant Anal.* 19: 1119–1131.

Noble, A.D., M.V. Fey, and M.E. Sumner. 1988a. Calcium aluminum balance and the growth of soybean roots in nutrient solutions. *Soil Sci. Soc. Am. J.* 52:1651–1656.

Noble, A.D., M.E. Sumner, and A.K. Alva. 1988b. The pH dependency of aluminum phytotoxicity: Alleviation by calcium sulfate. *Soil Sci. Soc. Am. J.* 52:1398–1402.

Nosko, P., P. Brassard, J.R. Kramer, and K.A. Kershaw. 1988. The effect of aluminum on seed germination and early seedling establishment growth and respiration of white spruce (*Picea glauca*). *Can. In. Bot.* 66:2305–2310.

Oakes, A.J., and C.D. Foy. 1984. Acid soil tolerance of *Leucaena* species in greenhouse trials. *J. Plant Nutr.* 7:1759–1774.

O'Brien, L.O., and M.E. Sumner. 1988. Effects of phosphogypsum on leachate and soil chemical composition. *Commun. Soil Sci. Plant Anal.* 19:1319–1329.

Parker, D.R., T.B. Kinraide, and L.W. Zelazny. 1988. Aluminum speciation and phytotoxicity in dilute hydroxy aluminum solutions. *Soil Sci. Soc. Am. J.* 52:438–444.

Parker, D.R., T.B. Kinraide, and L.W. Zelazny. 1989. On the phytotoxicity of polynuclear hydroxy Al complexes. *Soil Sci. Soc. Am. J.* 53:789–796.

Parrot, W.A., and L.H. Bouton. 1990. Aluminum tolerance in alfalfa as expressed in tissue culture. *Crop Sci.* 30:387–389.

Paulino, V.T., J. Olivares, and E.J. Bedmar. 1987. Nodulation and nitrogenase activity of pea nodules as affected by low pH and aluminum. *Plant Soil* 101:299–302.

Pearson, R.W., J. Childs, and Z.F. Lund. 1973. Uniformity of limestone mixing in acid subsoil as a factor in cotton root penetration. *Soil Sci. Soc. Am. Proc.* 37:727–732.

Pearson, R.W., F. Abruna, and J. Vicente-Chandler. 1962. Effect of lime and nitrogen application on downward movement of calcium and magnesium in two humid tropical soils of Puerto Rico. *Soil Sci.* 93:77–82.

Pessarakli, M., J.T. Huber, and T.C. Jackson. 1989a. Protein synthesis in green beans under salt stress with two nitrogen sources. *J. Plant Nutr.* 12:1361–1377.

Pessarakli, M., J.T. Huber, and T.C. Tucker. 1989b. Protein synthesis in green beans under salt stress conditions. *J. Plant Nutr.* 12:1105–1121.

Plucknett, D.L. 1972. The use of soluble silicates in Hawaiian agriculture. *Dept. of Agric. Papers*, Queensland University, St. Lucia, Australia, Vol. 1, No. 67:201–223.

Postel, S. 1990. Saving water in agriculture. *In* L.R. Brown, A. Durning, C. Flavin, H. French, J. Jacobson, M. Lowe, et al. (Eds.) *State of the World.* W.W. Norton and Co., New York, pp. 39–58.

Radcliffe, D.E., R.L. Clark, and M.E. Sumner. 1986. Effect of gypsum and deep rooting perennial on subsoil mechanical impedance. *Soil Sci. Soc. Am. J.* 50:1566–1570.

Ramirez, R., and T. Berensel. 1984. Effect of aluminum on the accumulation and distribution of P, Ca, Mg, and Al in maize lines. *Agronomia Tropical* 34:143–152.

Ramirez, R., H.H. Rodriguez, and J.E. Williams. 1986. Relationships between the root cation exchange capacity of sorghum and the tolerance to aluminum toxicity. *Agronomia Tropical* 36:115–128.

Rashid, A., F.M. Chaudhry, and M. Sharif. 1976. Micronutrient availability to cereals from calcareous soil. III. Zinc absorption by rice and its inhibition by important ions of submerged soils. *Plant Soil* 45:613–623.

Rechcigl, J.E., K.L. Edminsten, D.D. Wolf, and R.B. Reneau, Jr. 1987. Influence of calcium, nitrogen, and pH on alfalfa root growth and nitrogen fixation using the implanted soil mass technique. *Agron. J.* 79:926–928.

Rechcigl, J.E., K.L. Edminsten, D.D. Wolf, and R.B. Reneau, Jr. 1988. Response of alfalfa grown on acid soil to different chemical amendments. *Agron. J.* 80:515–518.

Rechcigl, J.E., R.B. Reneau, Jr., and D.E. Starner. 1985a. Effect of subsurface amendments and irrigation on alfalfa growth. *Agron. J.* 77:72–75.

Rechcigl, J.E., R.B. Reneau, Jr., and D.E. Starner. 1985b. Movement of subsurface fertilizer and limestone under irrigated and non-irrigated conditions. *Soil Sci.* 140:442–448.

Rechcigl, J.E., D.D. Wolk, R.B. Reneau, Jr., and W. Kroontje. 1985c. Influence of surface liming on the yield and nutrient concentration of alfalfa established using no tillage techniques. *Agron. J.* 77:956–959.

Reddy, M.R. 1987. Macronutrients. *Soybean Genetics Newsletter* 14:232–234.

Reddy, M.R., and S.J. Dunn. 1987. Differential response of soybean genotypes to soil pH and manganese. *Plant Soil* 101:123–126.

Rengel, Z., and D.L. Robinson. 1989. Aluminum effects on micronutrient intake by annual rye grass. *Commun. Soil Sci. Plant Anal.* 20:253–270.

Richardson, A.E., M.A. Djordjevic, B.G. Rolfe, and R.J. Simpson. 1988a. Effects of pH, Ca, and Al on the exudation from clover seedlings of compounds that induce the expression of nodulation genes in *Rhizobium trifolii*. *Plant Soil* 109:37–47.

Richardson, A.E., R.J. Simpson, M.A. Djordjevic, and B.G. Rolfe. 1988b. Expression of nodulation genes in *Rhizobium leguminosarum biovar trifolii* is affected by low pH and by Ca and Al ions. *Appl. Environ. Microbiol.* 54: 2541–2548.

Ritchey, K.D., V.C. Baligar, and R. Wright. 1988a. Wheat seedling responses to soil acidity and implications for subsoil rooting. *Commun. Soil Sci. Plant Anal.* 19:1285–1293.

Ritchey, K.D., J.E. Da Silva, and V.F. Costa. 1982. Calcium deficiency in clayey B horizons of savannah oxisols. *Soil Sci.* 133:378–382.

Ritchey, K.D., D.M.G. Sousa, and G.C. Rodriguez. 1988b. Biological tests for diagnosis of calcium deficiency and aluminum toxicity in seeds. *Revista Brasileira De Ciencia Do Solo* 12:113–120.

Ritchey, K.D., J.E. Da Silva, W. Espinoza and E. Lobato. 1981. Leaching of calcium sulfate to improve rooting in aluminum toxic, calcium deficient oxisol B horizons. *Leaching and Subsoil Utilization Newsletter*. EBRAPA, Planaltina, D.F., Brazil. Also in *Agron. Abstr.* 1981.

Ritchie, F.T. 1987. Efficient use of precipitation in humid regions. *In* L.L. Boersma, D.E. Elrick, R.B. Corey, H.H. Cheng, T.C. Tucker, E.M. Rutteds (Eds.). *Future Developments in Soil Science Research*. Soil Sci. Soc. Am., Madison, Wisconsin, pp. 327–336.

Robson, A.D., and J.F. Loneragan. 1970. Sensitivity of annual *Medicago* species to manganese toxicity as affected by Ca and pH. *Aust. J. Agric. Res.* 21:223–232.

Romer, W., J. Augustin, and G. Schilling. 1989. The relationship between phosphate absorption and root length in nine wheat cultivars. *In* B.C. Loughman, O. Gasparikova, and J. Kolek. (Eds.). *Structural and Functional Aspects of Transport in Roots*. Kluwer Academic Publishers, pp. 123–125. Boston, Dordrecht.

Rorison, I.H., and D. Robinson. 1984. Calcium as an environmental variable. *Plant Cell Environ.* 7:381–390.

Rost-Siebert, K. 1983. Aluminum toxizitat und tateranz an keinpflanzen von fichte (*Picea abies* Karst) and buche (*Fagus silvaticus* L. *Allg. Forstztz* 26/27:686–689.

Roy, A.K., A. Sharma, and G. Talukder. 1988. Some aspects of aluminum toxicity in plants. *Bot. Rev.* 54:145–177.

Salinas, J.G., and P.A. Sanchez. 1976. Soil plant relationships affecting varietal and species differences in tolerance to low available phosphorus. *Ciencia E. Cultura* 28:156–168.

Sanchez, P.A., and J.G. Salinas. 1981. Low input technologies for managing oxisols and ultisols in tropical America. *Adv. Agron.* 34:279–406.

Saneoka, H., N. Kanda, and S. Ogata. 1986. Differential tolerance among tropical forage crops to problem soil conditions. 1. The effects of low pH and aluminum in the culture medium on growth and nutrient uptake of several tropical forage crops. *J. Jap. Soc. Grassl. Sci.* 32:251–260.

Sangster, A.G., M.J. Hodson, and D.M. Parry. 1983. Silicon deposition and anatomical studies in the inflorescence bract of four *Phalaris* species with their possible relevance to carcinogens. *New Phytol.* 93:105–122.

Sedberry, J.E. Jr., P.G. Shilling, F.E. Wilson, and F.J. Peterson. 1978. Diagnosis and correction of zinc problems in rice production. *Louisiana Agric. Exp. Stn. Bull.*, No. 708.

Sewell, M.C., and P.L. Gainey. 1930. Interrelation of nutrients and soil reaction on growth and inoculation of alfalfa. *Soil Sci.* 30:297–306.

Shokohiford, G., K. Sakagami, R. Hamada, and H. Matsumoto. 1989. Effect of amending materials on growth of radish plant in salinized soil. *J. Plant Nutr.* 12:1195–1214.

Shortle, W.C., and K.T. Smith. 1988. Aluminum induced calcium deficiency syndrome in declining red spruce. *Science* 240:1017–1018.

Shuman, L.M., E.L. Ramseur, and R.R. Duncan. 1990. Soil aluminum effects on the growth and alumimun concentration of sorghum. *Agron. J.* 82:313–318.

Simpson, J.R., and J. Lipsett. 1973. Effects of surface moisture supply on the subsoil nutritional requirements of lucerne (*Medicago sativa* L.). *Aust. J. Agric. Res.* 24:199–209.

Simpson, J.R., and A. Pinkerton. 1978. Effects of subsoil acidity and liming on the root behavior and shoot growth of some leguminous plants. *Inf. Ser. New Zeal. Dept. Sci. Ind. Res.* 134 (Plant Nutrition V) 2:449–453.

Simpson, J.R., A. Pinkerton, and T. Lazdovskis. 1977. Effects of subsoil calcium on the root growth of some lucerne genotypes (*Medicago sativa* L.) in acidic soil profiles. *Aust. J. Agric. Res.* 28:629–638.

Simpson, J.R., A. Pinkerton, and J. Lazdovskis. 1979. Interacting effects of subsoil acidity and water on the root behavior and shoot growth of four genotypes of lucerne, *Medicago sativa. Aust. J. Agric. Res.* 30:609–620.

Sims, J.L., and K.L. Wells. 1985. Managing acid soils for production of burley tobacco. *University Kentucky Coop. Ext. Ser. Coll. of Agric.* AGR 109, Lexington, 3 pp.

Singleton, P.L., D.C. Edwards, E.E. Smart, and D.M. Wheeler. 1987. Soil acidity and aluminum and manganese toxicity on the TE Kauwhata Area, North Island, New Zealand. *New Zeal. Agric. Res.* 30:517–528.

Smit. H.P., W.G. Keltjens, and N. Van Breeman. 1987. Effects of soil acidity on Douglas fir seedlings. 2. The role of pH, aluminum concentration, and nitrogen nutrition (pot experiment). *Neth. J. Agric. Sci.* 35:537–540.

Smith, G.S., D.C. Edmeades, and M. Upsdell. 1987. Manganese status of New Zealand pastures. 1. Toxicity in ryegrass, white clover, and lucerne. *New Zeal. Agric. Res.* 26:215–221.

Snyder, G.H., E.O. Burt, and G.J. Gascho. 1979. Correcting pH induced manganese deficiency in Bermuda grass turf. *Agron. J.* 71:603–608.

Snyder, G.H., D.B. Jones, and G.J. Gascho. 1986. Silica fertilization of rice on Everglades histosols. *Soil Sci. Soc. Am. J.* 50:1259–1263.

Soileau, J.M., and O.P. Engelstad. 1969. Cotton growth in an acid, fragipan subsoil. I. Effects of physical soil properties, liming, and fertilization on root penetration. *Soil Sci. Soc. Am. Proc.* 33:915–919.

Soileau, J.M., O.P. Engelstad, and J.B. Martin, Jr. 1969. Cotton growth in an acid, fragipan subsoil. II. Effects of soluble calcium, magnesium, and aluminum on roots and tops. *Soil Sci. Soc. Am. Proc.* 33:919–924.

Soon, Y.K. 1988. Interactions between calcium amendment and phosphate on the responses of alfalfa and barley growth on acid soils. *Commun. Soil Sci. Plant Anal.* 19:1343–1353.

Spurway, C.H. 1941. Soil reaction (pH) preference of plants. *Michigan Agric. Exp. Stn. Bull.*, No. 306:1–36.

Stassart, J.L. 1987. Ionic composition and metabolic changes in salt stressed roots. *Developments in plant sciences*, Vol. 36:247–249.

Subbiah, C.C., P. Manikandan, and Y. Joshi. 1986. Yellow leaf spot of cashew *Anacardium occidentale*. A case of molybdenum deficiency. *Plant Soil* 94: 35–42.

Sumner, M.E., and E. Carter. 1987. Amelioration of subsoil acidity. *Commun. Soil Sci. Plant Anal.* 19:1309–1318.

Sumner, M.E., H. Shahanelek, J. Bouton, and J. Hammel. 1986. Amelioration of an acid soil profile through deep plowing and surface application of gypsum. *Soil Sci. Soc. Am. J.* 50:1254–1258.

Sundram, J. 1979. Studies on the effect of pH on *Rhizobium*-legume association. *Soil Microbiology and Plant Nutrition Symp.*, Kuala Lampur, 1976.

Suresh, R., C.D. Foy, and J.R. Weidner. 1989. Effects of water deficit and manganese toxicity on two cultivars of soybeans (*Glycine max* L. Merr.). Possible applications in remote sensing. *J. Plant Nutr.* 12:995–1003.

Takahashi, E., and Y. Miyake. 1976. Silica accumulation in the different growth states and organs of plants. Comparative studies on the silica nutrition in plants. (Part 8). *Abstr. J. Sci. Soil Man.*, Vol. 47, No. 8:338–341.

Takahashi, E., and Y. Miyake. 1982. The effects of silicon on the growth of cucumber plant. Comparative studies on the silica nutrition. *In* M. Scaife (Ed.) Plant Nutrition 1982. Proc. Ninth Int. Plant Nutrition Colloquium, Warwick Univ., England. Vol. 2, pp. 664–669. Commonwealth Bureau, Farnham House, Farnham Royal, Slough SL2 3BN, U.K.

Taylor, G.J. 1988. The physiology of aluminum tolerance in higher plants. *Commum. Soil Sci. Plant Annal.* 19:1179–1194.

Taylor, R.W., I.O. Ibeabuchi, and J.W. Shuford. 1989. Growth and metal accumulation of forage grasses at various clipping dates on an acid mine spoil. *J. Environ. Sci. Health*, A24:2, pp. 195–204.

Thaworuwong, N., and A. van Diest. 1975. Influence of high acidity and aluminum on the growth of lowland rice. *Plant Soil* 41:141–159.

Thornton, F.C., and C.B. Davy. 1983. Acid tolerance of *Rhizobium trifolii* in culture media. *Soil Sci. Soc. Am. J.* 47:496–501.

Traina, S.J., G. Sposito, D. Hesterberg, and V. Kafhafi. 1986. Effects of pH and organic acids on orthophosphate solubility in an acid montmorillonite soil. *Soil Sci. Soc. Am. J.* 50:45–52.

Tupper, G.R., H.B. Pringle III, M.W. Ebelhar, and J.G. Aneville. 1987. Soybean yield and economic response to broadcast, incorporated, and deep band placement of lime on low pH soils. *Bull. Mississippi Agric. For. Expt. Stn.*, No. 950, 7 pp.

Wallace, A. 1989. Relationships among nitrogen, silicon, and heavy metal uptake by plants. *Soil Sci.* 147:457–460.

Wallace, A., E. Frohlich, and A.R. Lunt. 1966. Calcium requirements of higher plants. *Nature* 209:634.

Wersuhn, G., H.H. Nhi, E. Tellhelm, and G. Reinke. 1988. Aluminum tolerant regenerants from potato cell cultures. *Potato Research* 31:305–310.

Wilkins, D.A., and M.J. Hodson. 1989. The effects of aluminum and *Poxillus involutus* Fr. on the growth of Norway spruce, *Picea abies* L. Karst. *New Phytol.* 113:225–232.

Wilkinson, R.E., and R.R. Duncan. 1989. H^+, Ca^{++}, and Mn^{++} influence on sorghum seedling shoot growth. *J. Plant Nutr.* 12:1395–1407.

Wood, M., and J.E. Cooper. 1988. Acidity-aluminum and multiplication of *Rhizobium trifolii*. Possible mechanisms of aluminum toxicity. *Soil Biol. Biochem.* 20:95–100.

Wright, R.J. 1989. Soil aluminum toxicity and plant growth. *Commun. Soil Sci. Plant Anal.* 20:1479–1497.

Wright, R.J., V.C. Baligar, and S.F. Wright. 1987. The influence of acid soil factors on the growth of snap beans in major Appalachian soils. *Commun. Soil Sci. Plant Anal.* 18:1235–1252.

Wright, R.J., V.C. Baligar, K.D. Ritchey, and S.F. Wright. 1989. Influence of soil solution aluminum on root elongation of wheat seedlings. *Plant Soil* 113:294–298.

Wright, R.J., J.L. Hern, V.C. Baligar, and O.L. Bennett. 1985. The effect of surface applied soil amendments on barley root growth in an acid subsoil. *Commun. Soil Sci. Plant Anal.* 16:179–192.

Yost, R.S., and R.L. Fox. 1979. Contribution of mycorrhizae to the P nutrition of crops growing on an oxisol. *Agron. J.* 71:903–908.

Root Extraction of Nutrients Associated with Long-Term Soil Management

A.N. Sharpley, J.J. Meisinger, J.F. Power, and D.L. Suarez

I. Introduction 151
II. Effects of Long-Term Soil Management on Root Extraction of Nutrients 152
III. Interaction Among Management Factors 189
IV. Modeling Long-Term Management Effects on Root Extraction of Nutrients 192
V. Conclusions 199
References 200

I. Introduction

A major charge to agricultural research is to identify management practices that maintain long-term soil fertility and crop production with reduced chemical inputs and tillage operations. Of prime importance is gaining a fundamental understanding of soil physical, chemical, and biological factors limiting crop production. In general, soil water content of the root zone is the major limitation to root growth, but many other dynamic variables also affect the root extraction of nutrients.

Management alternatives to maximize the root extraction of nutrients will differ between nutrient limiting and sufficient or excess sites. For example, under conditions where essential plant nutrients are limiting, application of fertilizer and other soil amendments, such as lime and animal composts, will be required. In contrast, at sites where plant nutrients are present in sufficient, or even excess, amounts, management practices should maximize utilization of residual nutrients by coinciding

Advances in Soil Science, Volume 19

zones of active root growth and nutrient accumulation. In addition, management alternatives for these different sites must consider both farm profitability and environmental quality aspects; in some cases, compromises may be necessary.

Constraints on the root extraction of nutrients associated with long-term soil management can be divided into non-manageable and manageable variables. Non-manageable variables include climatic, edaphic, and economic limitations. Climatic conditions may limit root extraction of nutrients through the effect of temperature and precipitation on evapotranspiration (ET), organic matter decomposition, and soil biological activity. Edaphic variables including soil depth (shallow), texture (coarse sands or heavy clays), and topography (steep), may exert physical limitations. In addition, market prices and government commodity programs may dictate selection of a crop or rotation that may not utilize available soil nutrients most efficiently. These limitations may be ameliorated by the identification and utilization of manageable variables: physical and chemical modification of the root environment; soil organic matter, through residue and manure applications; fertilizer amendments; irrigation; salinity; indigenous and introduced earthworm and mycorrhizal populations; soil conservation; and cropping systems.

The study of the long-term effects of agricultural management on the root extraction of nutrients is obviously time consuming and can be expensive, labor intensive, and site specific. Consequently, there is an immediate need for the development and application of models capable of simulating the effect of agricultural management on soil physical, chemical, and biological properties and on plant nutrient uptake and yields. At best, these models may provide accurate estimates of the relative management effects on nutrient uptake and identify potential alternative agricultural management systems that are both profitable and environmentally sound.

This paper evaluates the current state of knowledge on utilizing manageable variables to maximize the root extraction of nutrients and identifies associated short- and long-term research needs. Examples of interactions among the manageable variables are given, highlighting the dynamic nature of the complex processes involved. Several types of models simulating the root extraction of nutrients are discussed in terms of what they *can* and *cannot* do. In this article, major emphasis is given to the plant nutrients nitrogen (N) and phosphorus (P).

II. Effects of Long-Term Soil Management on Root Extraction of Nutrients

A. Effects of Root Environment Modification

Modifying the root environment to increase root growth is the most direct method of changing the root extraction of nutrients. Modification could increase rooting depth or density within the root zone. Several root environmental changes have been highly successful, such as drainage, salinity control, reducing mechanical impedance, and increasing soil pH.

1. Drainage

Roots of most crop plants require oxygen; if the oxygen supply is stopped, root elongation can stop within a few minutes (Huck, 1970). A healthy root system requires ready access to oxygen since nutrient absorption is governed by the rate of root respiration. Oxygen diffuses through air 10^4 times faster than through water; thus, a very wet soil will have a greatly diminished oxygen transport capacity and the roots will have a corresponding reduction in nutrient uptake capacity. Drainage of excess water through surface or subsurface mechanisms is an important factor to increase root growth and nutrient uptake. Drainage is also essential to control salinity and thereby increase root growth in irrigated agriculture (Van Schilfgaarde, 1974; Taylor, 1983; Payne and Gregory, 1988).

Excess water can cause rapid depletion of soil oxygen, depending on the demand for oxygen and texture. For example, a typical loamy sand will contain about 30 L of oxygen m^{-3} after free drainage of water, while a clay soil will contain about $5\,L\,m^{-3}$. Soil oxygen consumption rates have been reported as high as $17\,L\,m^{-3}\,day^{-1}$ under summer conditions and about one-tenth this value in winter (Currie, 1970). Therefore, the oxygen content of a soil can be greatly reduced in less than two days depending on soil texture and temperature (Van Schilfgaarde, 1974; Cannell and Jackson, 1981). Replenishing soil oxygen depends on the soil oxygen diffusion rate which has been shown to be directly related to root growth (Stolzy et al., 1961; Williamson, 1964; Grable, 1966; Chaudhary et al., 1975).

Drainage can impact root and plant growth by changing the water table height with subsurface drainage or by changing the number of days of surface flooding with surface drainage techniques. The impact of these two variables on corn was studied by Chaudhary et al. (1975) using field lysimeters to control the water regime. This study (Table 1) shows that the optimum height of the water table varied with climate; a high water table improved corn (*Zea mays*) growth in a dry year but a lower water table improved growth in a wet year. The impact of surface ponding is also shown and indicates that corn suffered a 10% growth decline with as little as two days of flooding occurring once a year. Other research has shown that crop species differ in their sensitivity to aeration stress with small grains being more tolerant than corn (Chaudhary et al., 1974), while corn was more tolerant than soybean (*Glycine max*) (De Boer and Ritter, 1970). Others have shown significant variation among cultivars toward aeration stress in corn (Fausey et al., 1983), white bean (*Phaseolus vulgaris* L) (Schumacher and Smucker, 1984), and soybean (Stone, 1985).

Table 1. Response of corn total dry matter production (as percentage of best treatment) to various depths of water table and various surface flooding treatments (data adapted from Chaudhary et al., 1975).

Water Table Depth	Relative Yield		Surface Flooding Duration	Flooding Cycle	
	1971 (wet yr.)	1972 (dry yr.)		every 14/28 da.	one time
cm	%		days	%	
60	70	100	0	100	100
90	87	98	1	92	98
120	100	84	2	82	91
150	84	76	3	64	82
—	—	—	4	46	72

The effects of drainage on the root extraction of nutrients is complicated by the fact that drainage affects oxygen supply, plant pathogens, plant hormones, soil temperature, available water, and nutrient losses (especially N). The only general conclusion that can be drawn is that surface and/or sub-surface drainage offers a direct and readily achievable way to increase the root extraction of nutrients to the extent that it increases crop dry matter production.

2. Salinity

Under conditions of extensive salinity damage, fertilizer requirements can likely be reduced in relation to the crop yield. For example, if 30% yield reductions are predicted based on salinity determinations, then N fertilization could be reduced by at least 30%, as the total crop fertilizer requirements are usually based on biomass production.

In most instances, it is preferable to switch to a more salt tolerant crop rather than suffer high yield losses due to salinity. Under irrigated conditions with high salinity water, the leaching fraction (percent of applied water leached) can be adjusted to some extent to decrease soil salinity. Generally, increasing the leaching fraction beyond 50% of the applied water is not feasible. In many instances, the attainable leaching fraction may be insufficient due to inadequate soil infiltration. Often some improvement is possible by management practices (discussed elsewhere in this chapter) such as increasing crop residues, avoiding surface compaction, etc. When salt damage is associated with inadequate drainage, installation of subsurface drainage, with the subsequent leaching of the soil profile, is recommended.

Under non-irrigated conditions, development of saline soils is often associated with excessive deep percolation, high water tables, and the upward movement of water and salts to the surface in low areas of the field. Changes in cropping patterns (for example from wheat [*Triticum vulgare*] to alfalfa [*Medicago sativa*]) to increase water utilization are often beneficial. Increased water use lowers the water table and restores the downward movement of water and salts.

3. Mechanical Impedance

Mechanical impedance refers to the mechanical resistance of a volume of soil to root penetration. High mechanical impedance can be caused by natural fragipans, by compaction during tillage such as plow pans, by lost soil aggregation due to excess tillage or organic matter loss, or by hardness caused by loss of water (Bowen, 1981). Roots grow as new cells behind the root tip, take up water, and expand against the constraining forces of the cell wall and the surrounding soil. The difference between the cell's turgor pressure and the cell wall constraining pressure is termed the root growth pressure (Gill and Bolt, 1955). Root growth pressures of several crops range from 700 to 2500 kPa (Gill and Bolt, 1955; Stolzy and Barley, 1968; Taylor and Ratliff, 1969). Soil mechanical impedance is difficult to measure due to changes in moisture, the effects of soil texture, and soil aggregation. The most common method is with a penetrometer reading calibrated to plant response. Soil bulk density also gives a rough estimate, with high bulk densities related to reduced root growth (Veihmeyer and Hendrickson, 1948). A general rule of thumb is that bulk densities above 1.55, 1.65, 1.80, and 1.85 $g\,cm^{-3}$ will severely impede roots on clay loams, silt loams, fine sandy loams, and loamy fine sands, respectively (Bowen, 1981; Jones, 1983).

Many studies have shown that high soil impedance reduces root growth. Taylor and Ratliff (1969) used laboratory compacted soil to study the root elongation of cotton (*Gossypium hirsutum*) and peanut (*Arachis hypogae*). They found marked differences between these crops (Fig. 1); cotton root elongation was reduced 50% at 700 kPa while the corresponding value for peanuts was at about 2000 kPa. Phillips and Kirkham (1962) showed that soil compaction by tractor traffic significantly increased bulk density, reduced air permeability, reduced corn root weights, and decreased corn yields 15–40%. On a Norfolk loamy (siliceous, thermic typic Paleudult) sand in Alabama, seed cotton yields were reduced about 60%, 75%, and 80% when a dense soil layer measuring 800 kPa was encountered by roots at 30, 20, or 10 cm below the soil surface, respectively (Lowry et al., 1970).

Modifying high mechanical impedance usually involves tillage or crop selection. The data of Figure 1 indicate that peanuts should be able to tolerate high mechanical impedance better than cotton. Attempts to improve rooting with tillage have been mixed and highly variable. For example, Van Doren and Haynes (1961) observed no yield increase with corn, oats (*Avena sativa*), or clover as a result of subsoiling or chiseling Ohio soils to 38 or 46 cm. On the other hand, cotton growing on hardpan

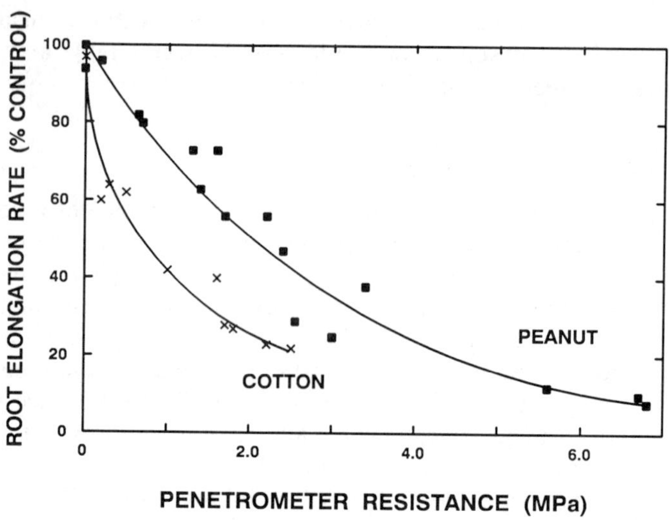

Figure 1. Relationship between root elongation and penetrometer resistance for peanut and cotton as reported by Taylor and Ratliff (1969) and summarized by Gregory (1988).

soils in Louisiana benefited from deep tillage as a result of increased infiltration, particularly in dry years (Saveson et al., 1961). The benefits of modifying high impedance layers depend on several factors: severity and nature of the problem, climate conditions (water stress), crop grown, and general soil fertility. The reviews by Unger et al. (1981) and Reicosky (1983) consider these interacting factors in greater detail.

4. Soil pH

As soils develop in humid climates, they naturally become acidic as bases are leached out with percolating water. Many crops exhibit reduced root growth when encountering a low soil pH. These reductions are most commonly associated with increases in Al (Fig. 2), but can also be related to increased Mn or deficiencies of Ca, Mg, or Mo (Taylor, 1983; Foy, 1984). Roots exhibit damage from Al by a reduction in elongation, probably associated with the alteration of cell membranes, and a reduction of cell division, probably associated with reduced DNA synthesis or with Ca deficiency (Ulmer, 1979; Hecht-Buchholz and Foy, 1981; Foy, 1984) Aluminum injured roots are stubby, lack fine branching, and have areas of brown cortical tissue (Foy, 1974).

Improving the root extraction of nutrients in soils or subsoils generally involves neutralizing the acid soil with liming materials, growing tolerant crops, or growing Al tolerant cultivars. The use of liming materials to increase soil pH and root growth is an old practice which has been

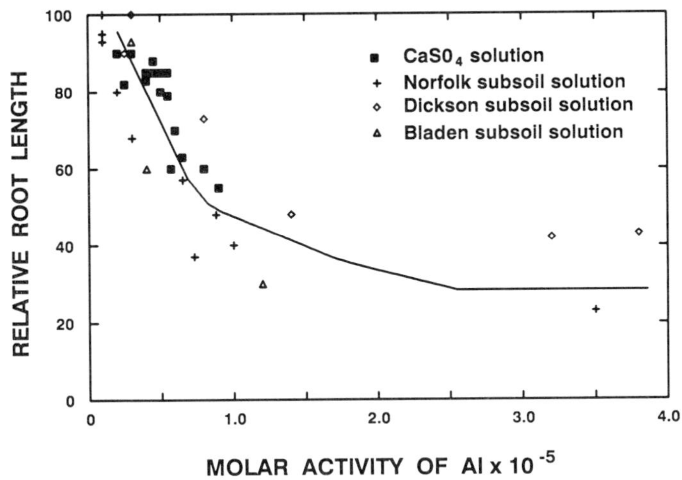

Figure 2. Influence of aluminum activity on root growth (Adams and Lund, 1966. From effect of chemical activity of soil solution aluminum on cotton root penetration of acid subsoils, *Soil Science* 101:193–198, © by Williams and Wilkins, 1966).

adequately reviewed by several authors (Adams, 1984; Kamprath, 1984; McLean and Brown, 1984). As their reviews indicate, long-term soil management that maintains surface soil pH with an active liming program is essential for the satisfactory root extraction of nutrients. Reducing subsoil acidity has not been highly successful due to the physical limitations of mixing lime into subsoils (Reicosky, 1983). However, surface applications of gypsum, with sufficient time for transport into the subsoil, have been shown to increase crop yields, increase subsoil Ca, and decrease exchangeable Al (Hammel et al., 1985; Sumner et al., 1986; Farina and Channon, 1988). The mechanisms for the beneficial effect of gypsum have not been clearly defined but likely involve ion-pair formation (Singh, 1984; Sumner et al., 1986; Nobel et al., 1988) or formation of aluminum-hydroxy-sulfate (Hue et al., 1985). The best method to avoid subsoil acidity is not to let it develop, i.e., maintain proper surface soil pH through an aggressive lime program (Adams, 1984). This is especially important in modern reduced tillage systems.

The extraction of nutrients in low pH soils can also be improved by switching to an acid tolerant crop, for example, substituting peanuts for cotton (Adams and Pearson, 1970), corn for sorghum (*Sorghum bicolor*) (Brenes and Pearson, 1973), or Bermuda grass (*Cynodon dactylon*) for fescue (*Festuca elatior*) (Adams, 1981). A general listing of crop tolerance to acidity is given by Adams (1981).

The root extraction of nutrients can also be increased by shifting to Al tolerant cultivars. Some crop cultivars showing a range of Al tolerance include alfalfa (Bouton et al., 1982), bird's-foot trefoil (*L. corniculatus*) (Alison and Hoveland, 1989), soybean (Arminger et al., 1968), wheat (Neenan, 1960), and corn (Lutz et al., 1971). Selecting a cultivar tolerant to a specific soil acidity problem is a very cost effective method of increasing nutrient utilization and farm profitability when dealing with subsoil acidity. Increasing the root extraction of nutrients in low pH soils or subsoils, therefore, requires a long-term liming program and the utilization of crops or cultivars which are tolerant of acid soil conditions.

At the other end of the pH scale, calcareous soils under cropped conditions in arid and semi-arid environments generally have pH values in situ of 6.7 to 7.5. These values are much lower than the commonly reported values of pH 8.0 to 8.6 based on determinations of a soil-water extract in the laboratory. The solubility of micronutrients Fe, Mn, Zn, Cu, and P decrease rapidly with increasing pH; however, the effect on total metal concentrations is not as severe since the majority of these nutrients are present in the solution as metal-organic ligand complexes (O'Connor et al., 1971).

At pH values above 7.5 to 8.0 (in situ), micronutrient problems, including P deficiency, can occur. In calcareous soils, high pH is associated with high carbonate alkalinity, which interferes with H^+ efflux pumping by the plant (Romheld and Marschner, 1983). The increased acidity and organic acids around the plant root increase micronutrient concentrations around the root, promoting adequate nutrient uptake.

Deficiencies in Fe and Zn under elevated pH conditions can be alleviated by the application of organic matter or micronutrient fertilizers (Mathers et al., 1980; Srivastava and Sethi, 1981). Under dry land conditions, some pH control is possible by the selection of N sources. Use of NH_4^+ results in net acidification of the soil and is suitable for helping to lower the pH of calcareous soils that may be above the normal range. Alternatively, the application of gypsum serves to precipitate carbonate as $CaCO_3$ and thus increases the Ca/HCO_3 ratio as well as lowering the pH of calcareous soils. In connection with this, the use of by-product gypsum to increase Ca availability and root growth in subsoils should be evaluated.

B. Effects of Organic Matter Management

1. Residue Management

Crop residue management practices may have significant effects on the root extraction of nutrients. Factors involved include quantity and type of crop residues used and time of residue incorporation. Quantity of residues involved may affect the amount of nutrients being cycled and potentially made available for root extraction after going through the mobilization-immobilization process. Increasing the crop residue rate can

Table 2. Uptake of N (kg ha^{-1}) from various sources at 1981 harvest of maize and soybean (whole plant) as affected by crop residue rate on the soil surface (from Power et al., 1986).

	Source of N				
Crop residue rate %	Crop residues	Residual fertilizer	Current fertilizer	Native soil N[+]	Total
Maize					
0	0	5	4	73	82
50	0	6	7	97	110
100	2	6	7	114	129
150	1	6	11	124	142
Soybean					
0	0	2	14	84	100
50	1	2	21	124	148
100	38	7	16	116	177
150	63	6	20	106	195

[+] For soybean, native soil N includes biologically fixed N.

also reduce evaporation losses and keep the surface soil moist, particularly if the residue is on the soil surface. This enhances microbial transformation of nutrients such as N, resulting in greater mineralization, availability, and uptake of indigenous soil N (Table 2).

The chemical composition of crop residues is particularly important in determining the balance between mobilization and immobilization processes. Generally, if the C:N ratio is greater than about 30 to 35, net immobilization of N in the residues results. With a ratio less than 25 to 30, net mobilization occurs. Similar thresholds for the ratio between residue C, P, S, and other plant nutrients can also be established. Power et al. (1986) demonstrated that little N in corn residues (wide C:N ratio) was mineralized and used by the next crop, whereas a large part of the N in soybean residues (narrow C:N ratio) was utilized by the next crop (Table 2).

Placement of crop residues is usually achieved through tillage. With no-till, crop residues are left on the soil surface; in contrast, clean tillage systems (plow or disk) largely incorporate residues. Many reduced and minimum tillage systems are intermediate to these extremes in terms of the degree of residue incorporation. When incorporated, the nutrients in the residues are distributed throughout the depth of tillage, in contrast to no-till systems where the nutrients are concentrated near the soil surface (Follett and Peterson, 1988). This distribution of nutrients may partially explain why Wilhelm et al. (1982) found greater wheat root growth near the surface of no-tilled compared to plowed soil. Also, the greater soil water content near the surface of no-tilled soils increases root activity as well as soil microorganism activity (Doran, 1980). These changes in microbial habitat and activity may have major effects on subsequent nutrient transformations and availability (Doran and Smith, 1987; Sharpley and Smith, 1989; Smith and Sharpley, 1990).

2. Manure Management

Manures and organic wastes are other sources of organic matter. As with crop residues, other factors such as composition, placement, timing, and rate influence nutrient cycling and availability from these sources. Animal manures vary widely in composition, water content, and other characteristics. If properly handled, they often serve as a major input of N, P, S, and other nutrients. The uses of animal manures and their potential contribution to the root extraction of nutrients by crops are discussed in Gilbertson et al. (1979). Rasmussen et al. (1989) and many others have documented the effects of manure on soil properties and crop production. If added in too large a quantity, animal manures may result in an excessive accumulation and leaching of nitrates beneath the crop root zone.

Other organic wastes that are of importance in localized regions include sewage sludge from municipalities, wastes from livestock slaughtering facilities, and wastes from the food processing and other industries. Again, composition, rate, placement, and time are major factors affecting nutrient availability.

C. Effects of Fertilizer Amendment

The type, placement, timing, and rate of fertilizer application must be carefully controlled to maximize nutrient availability and uptake while minimizing movement of mobile N below the root zone and stratification of nonmobile P in certain soil layers. The adoption of reduced tillage practices has further complicated and focused more attention on the ability of soil test procedures to reliably estimate fertilizer requirements. The major emphasis of this section is on P. More detailed information on the effect of agricultural management on soil N availability and fertilizer N use-efficiency is available from reviews by Halvorson et al. (1989), Peterson and Frye (1989), Power and Broadbent (1989), Russelle and Hargrove (1989), and Schepers and Fox (1989) and for sulfur by Stewart and Sharpley (1987) and Tabatabai (1984, 1986).

1. Fertilizer Type

Anhydrous ammonia accounts for the major proportion of fertilizer N use, due mainly to its lower cost than other forms. As a result of volatilization, immobilization, fixation, denitrification, and leaching losses, the use of slow release N fertilizers (e.g., sulfur-coated urea, urea formaldehyde compounds, etc.) has been evaluated under different climatic and management conditions. Generally, however, their cost has restricted widespread use. Nitrification inhibitors can be used to modify the proportion of nitrate to ammonium in the soil on a temporary basis, but the natural proportion of these N forms may be adequate depending on soil chemical and physical properties.

Slow release P fertilizers, such as P rock (PR), have also been evaluated as a source of available P over a growing season (Barnes and Kamprath, 1975; Khasawneh and Doll, 1978; Hammond et al., 1986; Hedley et al., 1990). Actual agronomic value is influenced by PR particle size, soil, climate, and plant management factors. Solubility of PR is generally restricted, however, to soils of low pH, low Ca, and P content. Even so, it may be possible to design agricultural management systems that take advantage of low cost PR. For example, in soils of neutral pH, it may be possible to apply a heavy initial dressing of finely ground PR and include a rotation of fine-rooted legumes to generate a low pH rhizosphere with low Ca concentrations and thus increase PR dissolution.

Other methods designed to increase acidity in the immediate PR environment and, thereby, its dissolution, include the addition of elemental S (Muchovej et al., 1989), ammonium forms of N fertilizer, or organic matter (Hedley et al., 1989). Research has shown that when N fertilizers are combined with PR by banding (Cabala-Rosand and Wild, 1982; Apthorp et al., 1987); granulating (Hedley et al., 1989), or compacting (Chien et al., 1987b), the dissolution of PR and plant uptake of P increase as well as providing plant N. The potential acidification of PR by composting with animal manure or organic residues may be enhanced by the chemical pretreatment of straw (L'honeux et al., 1988) or inoculation with fungi having lignocellulase activity and anaerobic N-fixing bacteria (Lynch and Wood, 1985). Further, the addition of mycorrhizal fungi to soils of low indigenous populations has been shown to increase P uptake from PR by wheat and field beans (*Phaseolus vulgaris*) (Kucey, 1987), ryegrass (*Lolium perenne* L.) (Powell, 1979), and sorghum (Salih et al., 1989). More research is needed, however, to evaluate the effectiveness of these amendments to enhance the root extraction of P for different soils and cropping practices.

Partial acidulation of PR (PAPR), specifically North Carolina PR (NCPR), can provide a viable economic and agronomic alternative to single (SSP) and triple super P (TSP). Chien and Friesen (1992) compared the effectiveness of various PR's in increasing maize-dry matter yields in greenhouse trials (Fig. 3). Assigning a value of 100% to TSP, the relative agronomic effectiveness of the various PR's ranged from 36% for Togo to 100% for ground (<0.075 mm) NCPR (Fig. 3). The relative agronomic effectiveness of unground and ground PR was a function of its reactivity as estimated by its solubility in neutral ammonium citrate (Fig. 3).

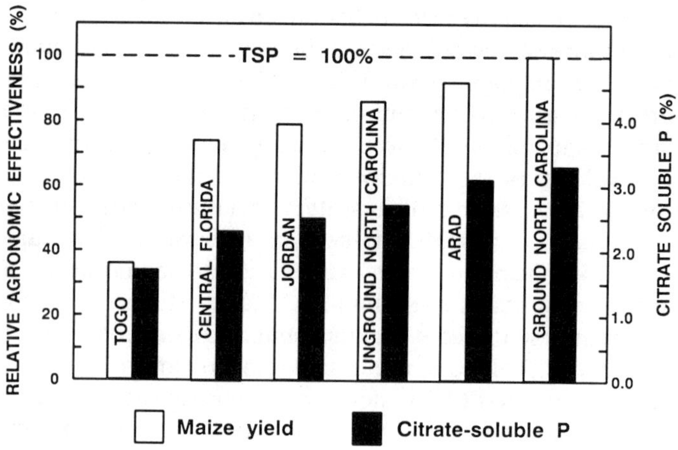

Figure 3. Relative agronomic effectiveness and citrate-soluble P of various phosphate rocks compared to triple super P (TSP) for maize (adapted from Chien and Friesen [1992] with the authors permission).

Research is needed to determine why PAPR's are sometimes more effective in supplying P and increasing yields than fully acidulated SSP or TSP (Misra et al., 1986; Marwaha, 1989; White et al., 1989). It is possible that the drop in pH in the soil surrounding a dissolving PAPR granule is less than that around a fully acidulated fertilizer (White et al., 1989). In acid soils, this could result in less dissolution of Fe and Al from soil minerals and subsequently less precipitation of insoluble Fe and Al phosphate reaction products. Residues in the PAPR granule and reaction products in the soil may, thus, sustain a higher solution P concentration than for SSP or TSP. It is also possible that the acidic solution produced by hydrolysis of the water soluble P component in the granule may dissolve some of the unacidulated rock residue. Further, the presence of water-soluble P in the PAPR granule may stimulate root growth around the granule, resulting in more efficient utilization of the PR residue.

Further research on the use of PR as an alternative P source is needed to evaluate cost benefit ratios as a function of edaphic and climatic conditions, material pretreatment, and application methods for different cropping systems (Chien and Friesen, 1992). The main use of PAPR's has been on acidic soils in tropical and New Zealand agricultural systems (Smyth and Sanchez, 1982; Hammond et al., 1986; Hedley et al., 1989a), where reduced fertilizer costs are of prime concern. An increasing adoption of "low-input" management and inclusion of forage legumes in United States crop rotations, along with the development of reactive PR sources, may increase the agronomic and economic benefit of PR. Thompson (1990) suggests that North Carolina reactive PR may have value for direct application, especially on moderately acid, medium to high organic matter soils low in available P and under forage legumes in low-input systems. It is also suggested that it may be the best P fertilizer available for organic farmers and others using only "natural" soil amendments.

The use of PR in forestry may also be a viable agronomic and economic alternative to soluble fertilizer P sources, where interactions between tree rhizosphere processes and PR may increase P availability and uptake efficiency (Gillespie and Pope, 1990). Rock P may also be a potential source of P in the revegetation of acid minespoils (Chien and Friesen, 1992). However, because of the acidic nature of these soils, can less-reactive PR sources be of agronomic value? What is the residual availability of PR, type of compounds formed, and rate of PR dissolution on minespoils? Research should evaluate the agronomic value of Ca from PR, as these soils are generally low in Ca and require liming.

Many questions remain; for example: could application of PR in wide bands alleviate slow dissolution rates, while limiting soil P fixation?; can PR grinding, which increases both cost and application problems, be eliminated for more reactive PR sources?; and can compaction of granulated PR with acid-forming N sources reduce the acidification needed and enhance PR dissolution and thereby its agronomic effectiveness? Research has demonstrated an increased agronomic effectiveness with 50:50 mixtures of PR's and TSP, which produced similar or higher yields of maize and wheat than TSP application to acid soils (Tiwari, 1979; Chien et al., 1987a). Subsequent research showed that compacted low-reactive PR and TSP were more effective in supplying P to maize than partial acidulation of the PR (Chien et al., 1987b; Menon and Chien, 1990). Although economically viable multinutrient fertilizer may be produced by compacting PR, TSP, urea, and potassium chloride, further agronomic evaluation of the use of such fertilizer products on acid, P-deficient soils is needed.

2. Fertilizer Placement

Potential immobilization of N by residues under no-till may increase broadcast fertilizer N requirements (Doran and Smith, 1987). Greater fertilizer N uptake by crops can, thus, be achieved with surface banding application or with subsurface knife applications. For example, unified N-P-K application increased grain sorghum yields relative to broadcast application to a greater extent in reduced, compared to conventional, tillage systems in Kansas (Lammond, 1987). Because of the lack of P mobility in the soil profile, fertilizer placement is generally more critical for P than N.

Several studies have shown a greater yield response to the banded application of fertilizer P at low rates, compared to broadcast or mixing

(Welch et al., 1966). In fact, Welch et al. (1966) observed greater P uptake and yield of corn with a combined banded (50%) and broadcast (50%) application ($40\,kg\,ha^{-1}$). The effect of P application, however, varies with soil and crop type. When P fertilizer is placed in a specific soil volume, root extraction of P depends on the rate of application, which affects soil P adsorption/desorption characteristics and diffusion, and on the stimulation of root growth in the fertilized soil volume. For six soils having a hundred-fold variation in P sorptivity, Holford (1989) found that fertilizer P effectiveness, as measured by the yield response in the first crop (wheat), residual effect in the second crop (clover), or cumulative recovery of applied P, was consistently greater for shallow banding at a 5-cm depth compared to banding at 15 cm and broadcast applications. The almost equal effect of mixing P throughout the soil regardless of P sorptivity, however, suggests that the important factor in maximizing fertilizer effectiveness is its positional availability in the root zone rather than the reduction of chemical immobilization by concentration in bands (Holford, 1989).

Positional availability will also be influenced by crop type. In order for banding or restricted fertilizer placement to increase the potential root extraction of P, the rate of P absorption and growth of roots in fertilized soil must increase to compensate for roots in unfertilized soil. Increased root growth in the P-fertilized volume of soil compared to unfertilized soil has been observed for corn (Anghioni and Barber, 1980), soybeans (Borkert and Barber, 1985), and wheat (Yao and Barber, 1986). In contrast, several studies have shown that flax does not respond to banded fertilizer due to an inability of its root system to expand and proliferate into and efficiently absorb P from high concentrations in the fertilized zone (Soper and Kalra, 1969; Strong and Soper, 1974). In the case of flax, increased P uptake and yield response was obtained when fortilizer P was placed 2 to 5 cm directly below the seed, ensuring adequate P levels during early growth (Bailey and Grant, 1989).

3. Fertilizer Timing

Due to the rapid adsorption of P by soil material, the timing of fertilizer application is more important for efficient N use than for P. Application of fertilizer N before or during periods when precipitation exceeds evapotranspiration must be evaluated in terms of crop N requirements and the risk of leaching. However, in some systems such as rain-fed winter wheat in Nebraska, partial leaching may be encouraged by the early application of N so that a major proportion of the stored water and N were available at similar depths in the root zone (Smika and Grabouski, 1976).

Crop utilization of N is usually greatest with split applications, where the time of N availability and crop uptake can be more closely synchronized. Although starter fertilizers may meet early season growth needs, yields may not be affected as the amount of N in starter fertilizers placed near the seed is only a small part of the total crop N requirement. Under irrigation, greatest N utilization by corn is usually obtained with side-dress application at the time of greatest N uptake. Fertigation may also facilitate easy application of N at appropriate times. Even so, the benefits of improved technology and information regarding fertilizer type, placement, and timing on fertilizer-use efficiency are, to a large extent, dependent on reliable soil test procedures. Advances in soil characterization, sampling, and fertilizer application technology may have made soil test methodology and interpretation the weakest link in the fertilizer recommendation process.

4. Soil Testing

At present, the main concerns regarding soil test procedures for fertilizer N, P, and S recommendations involve accounting for organic and residual soil nutrient contributions and sampling techniques for reduced or no-till management systems.

Nitrogen: Available soil N is comprised of N mineralized from soil organic matter during the growing season plus residual nitrate in the profile. Soil organic N mineralization can supply a significant proportion of a crop's N needs. In an attempt to quantify this input, Stanford and Smith (1972) developed a soil N mineralization potential (N_o) by laboratory incubation (12 to 30 weeks) and adjusted for field moisture and temperature. Its subsequent application has had some field success (Oyandel and Rodriguez, 1977; Smith et al., 1977). Recent modification of the N_o concept by using short-term mineralization, chemical extraction of N_o, and use of undisturbed soil are potential improvements of the test to quantify N_o (Cabrera and Kissel, 1988; Gianello and Bremner, 1988; Schepers and Saint Fort, 1988). Even so, no generally accepted index of soil N mineralization exists.

Estimates of residual nitrate in the root zone have been a useful index for dry land regions and for soils suspected of having an appreciable accumulation of nitrate in the profile. As these tests only measure the amount of nitrate present at the time of sampling, they have limited use as a soil N index under conditions conducive to leaching. Measurement of residual nitrate levels in surface soil samples (30 cm depth) collected when corn plants are 15 to 30 cm high (V6 stage of growth) is gaining acceptance as an available soil N test for supplemental or side-dress N applications in the northeast (Magdoff et al., 1984) and Iowa (Blackmer et al., 1989).

Plant tissue N tests may complement soil N tests, due to the fact that the plant root system will integrate physical, chemical, and biological factors contributing to the spatial variability of soil N supply power over a relatively large field volume. Although tissue testing procedures are routinely used for some fruit and nut crops, sugar beets (*Beta vulgaris*), and potatoes (*Solanum tuberosum*), their use on field crops such as cotton (Tucker and Murdock, 1984), corn, and soybean (McCleanhan and Killorn, 1988; Schepers and Below, 1988) is gaining acceptance.

Phosphorus: Organic P (P_o) can be an important source of plant-available P in some temperate soils and in soils under reduced tillage practices (Sharpley, 1985; Harrison, 1987). Consequently, it is possible that soil P fertility tests may be improved by accounting for or giving credit to readily mineralizable P_o as well as inorganic P. Daughtrey et al. (1973), Adepetu and Corey (1976), Bowman and Cole (1978), and Abbott (1987) found that potential soil P supply, as reflected by crop yields, was more closely estimated by including extractable P_o. Bowman and Cole (1978) used a modification of the Olsen bicarbonate test (Olsen et al., 1954), which measured the total amount of P (inorganic plus organic) extracted by the reagent. Where other soil P test methods are recommended, a similar adaptation may be used. A potential problem in estimating extractable P_o is that the conditions of P extraction may not represent conditions for P_o mineralization in the field. In addition, seasonal variations in climatic and soil conditions will have a greater effect on the amount of extractable P_o than inorganic P content. Consequently, caution must be exercised in relating the amounts of extractable P_o to expected crop response in the field.

Future research should also be directed towards the inclusion of quantity (concentration of sorbed P) and intensity (solution P concentration) factors in soil P test procedures (Moody et al., 1988; Kuo, 1990). In addition, the use of resin accumulators (Cooperband and Logan, 1991; Yang et al., 1991) or Fe oxide-impregnated paper strips (Menon et al., 1989, 1990) as a sink for plant available P are being evaluated as potential routine soil test methods. These improvements in soil P testing procedures will be of greater importance to conservation tillage practices, where residual fertilizer and crop residue P may accumulate near the soil surface (Triplett and Van Doren, 1969; Cruse et al., 1983; Rehm, 1992). As a result, adequate soil test P levels for reduced tillage may not be the same as for conventional tillage. Rehm (1992) suggests, from preliminary data, that higher soil test P levels may be required for optimum production in conservation tillage. There is, thus, a need to develop regional soil P test calibration and response data bases that encompass similar soils, crops, management, and climatic conditions.

Sulfur: Sulfur-deficiency diagnosis by soil testing methods have involved the extraction of soluble sulfate (S_i) salts with water, a weak electrolyte (0.01 *M* $CaCl_2$), or phosphate solutions (Walker and Doornenbal, 1972; Hamm et al., 1973; Maynard et al., 1982). These extractions, however, do not account for organic sulfur (S_o) mineralization. Consequently, deficient soil S levels may be indicated, although no crop response to S fertilization is obtained. Further complications arise from the fact that the contribution of S in wet and dry deposition (up to $50\,kg\,S\,ha^{-1}\,yr^{-1}$) can be of agronomic importance (Martin, 1980; Tabatabai, 1984; Morra and Dick, 1988).

Research is needed, therefore, to develop a soil test procedure which removes the most labile S_o fraction. As was the case for P_o, however, the conditions of chemical extraction are unlikely to simulate mineralization of S_o under field conditions. Nonlabile S_o may also slowly release large amounts of S_i relative to a small labile S_o pool. Improved soil S tests will probably involve the prediction of plant-available S dynamics and turnover under a wide range of environmental conditions using computer modeling. Refinements in methodology will also be needed before the role of inorganic and organic S in plant nutrition can be more comprehensively understood and reliable soil test procedures developed.

In summary, the following research needs and recommendations are identified concerning the effect of soil amendments on root extraction of nutrients:

a. Provide direct nitrification inhibitor and N stabilizer research to produce compounds and formulations geared to specific application and management techniques. The mode of action should be known so that chemical formulations can be made with predictable specificity and efficiency.
b. Evaluate the relative effectiveness of PAPR in association with inorganic and organic amendments to enhance the root extraction of P for different agricultural management systems as a function of soil, plant, and climatic factors.
c. Evaluate recent developments in equipment design regarding precise timing, distribution, injection, and placement of fertilizer material as an aid to maximize nutrient uptake and minimize losses due to N leaching and P stratification.
d. Tailor fertilizer application techniques to non-uniform soil areas rather than whole fields, giving credit for residual N and P effects from previous fertilization and crop residues and incorporating appropriate Geographic Information Systems (GIS).
e. Continue the development of soil and tissue N tests. Integrate these tests with mass balance, N fertilizer management approaches, and crop growth models to ensure timely application of N.
f. Integrate the existing knowledge on soil P dynamics into soil testing programs, utilizing nonconventional methods and user-friendly expert systems.
g. Integrate the microbial cycling of nutrients, fertilizer management practices, and crop nutrient requirements to improve fertilizer efficiency by synchronizing nutrient availability and crop requirements.

D. Effects of Irrigation

1. Influence of Irrigation Type

Sprinkler irrigation has little unique effect on the root extraction of nutrients. Processes and patterns are essentially the same as those encountered with rainfed agriculture. Because the quantity of water applied is much better controlled with sprinkler rather than gravity irrigation, the likelihood of nutrient leaching is lower (except on extremely sandy soils). In addition, sprinkling offers the opportunity to apply fertilizers in the irrigation water during the growing season to correct nutrient deficiencies.

There are several types of gravity irrigation systems, but with row crops the most common is furrow irrigation. A major problem associated with furrow irrigation is the uniformity of water application. In order to get enough water infiltration to recharge the soil water at the end of the furrow, water must pass over the soil at the beginning (top) of the furrow for a much greater length of time, creating an opportunity for considerable leaching. This is particularly apparent where the length of the irrigation furrow and soil infiltration rate are poorly matched (or often where the infiltration rate changes drastically across the length of the furrow). Surge irrigation techniques have been developed to partially alleviate this problem. With surge irrigation, water is rapidly and intermittently applied to the furrow. This results in less water being required to move water to the end of the furrow, better smoothing of the irrigation furrow, and a more controlled infiltration rate.

A unique situation arises with row crops from the common practice of running irrigation water between every other row. Fertilizers are often banded in the rows not receiving water. These practices create unique situations for water and nutrient movement, as well as root growth and nutrient utilization. However, adequate research information is not presently available to characterize this situation.

In addition to these common methods of irrigation, a number of specialized irrigation techniques are sometimes employed for unique situations. For example, with high-value crops, trickle and sub-irrigation may be utilized wherein water is applied through a network of porus tubes placed either on the soil surface or beneath the surface. Frequently, nutrients are injected directly into the water stream, thereby improving both water and fertilizer use efficiency. Ponding, flooding, or basin irrigation are also sometimes used. In these situations, excellent control of water is possible, thereby greatly reducing the possibility of the loss of soluble nutrients by leaching.

2. Influence of Irrigation Frequency

Irrigation water may be applied based on a number of different criteria. Often, irrigations are scheduled on the basis of climate (precipitation and

ET potential), crop development, soil "feel," or soil water monitoring devices (tensiometers, psychrometers, moisture blocks). The quantity and frequency of irrigation may also be controlled by unmanageable factors such as the availability and cost of irrigation water (for both pump and diversion systems). If significant plant water stress is permitted between irrigations, the uptake of mobile nutrients may be especially affected. For immobile nutrients such as P, plants appear able to take up and store sufficient nutrients during moist periods to meet the plant requirements for sometime thereafter (Power et al., 1961). It is not uncommon for irrigation water to contain significant quantities of soluble nutrients, particularly N. Credit must be given to N in irrigation water in determining fertilizer N requirements.

E. Effects of Salinity

Fertilization needs under saline conditions are generally considered to be comparable to those under non-saline conditions (Bernstein et al., 1974; Maas and Hoffman, 1977; Feigin, 1985). Plants respond favorably to additions of P fertilizer under saline conditions. However, Bernstein (1964) concluded that when salinity depressed yields by 50% or more (relative to non-saline and optimal fertility conditions), fertilization could be below the optimal levels without a major impact on yield. Adverse impacts due to salinity cannot be overcome by increased fertilization. However, the interactions between salinity and fertility are not well understood as there is seemingly contradictory information.

1. Nitrogen

Crops that demonstrate increased fertilizer requirements under saline conditions may be responding to changes in anion ratios in solution. Kafkafi et al. (1982) observed that tissue N was reduced in tomato as Cl concentration in the nutrient solution was increased. Similar effects have been found for potato (Saffigna and Keeney, 1977) and wheat (Torres and Bingham, 1973). Apparent increases in N uptake under saline conditions for crops such as corn may be due to a reduction in plant biomass and plant population, thus making more N available.

Changes in the apparent salt tolerance of crops have also been observed with N fertilization. Additions of N increased the sensitivity to salinity of corn and cotton (Khalil et al., 1967) as well as rice and wheat (Ogo and Morikawi, 1965). Similar results were obtained by Bernstein et al. (1974) for cabbage (*Brassica oleracea*), beets (*Beta vulgaris* L. cv. "Detroit Dark Red"), and broccoli (*Brassica oleracea* var. *capitata*). As observed with P fertilization, it seems that eliminating the fertilizer limitations to growth allows for salinity to become the growth-limiting factor. Thus, on experiments where fertilization is optimum, changes in crop growth with increase in salinity (sensitivity to salinity) are larger than on experiments where optimum fertilization was not achieved. Similarly, if a crop is N deficient under saline conditions, N fertilization may increase yield and this could be misconstrued as increased salt tolerance.

Most salinity–N studies have reported increases in crop yield with the addition of N. For example, Khalil et al. (1967) for corn and cotton, Dregne and Majallali (1969) for barley (*Hordeum vulgare*), Wagenet et al. (1983) for bean, and Ogo and Morikawi (1965) for wheat and rice, among others, all observed increased yield or growth with increased N applications under saline conditions. In these studies, optimal N fertilization under non-saline conditions was not evaluated. There are only a very few reports of increased yield under saline conditions after application of N above the optimum levels determined under non-saline conditions. It can be concluded that under saline conditions, N fertilization can be below the optimum levels established under non-saline conditions. Specific N fertilizer recommendations have to consider the extent of salinity damage as well as crop type (and often varietal differences).

2. Phosphorus

The majority of studies involving P-salinity interactions have been done in solution or sand culture under conditions that may not simulate the field situation. Bernstein et al. (1974) showed that, in a sand culture, corn, wheat, and vegetable crops all responded to P additions at low levels. However, under increasing salinity, P induced injury occurred at P levels that caused no injury at lower salinity levels.

Grattan and Maas (1984) observed P toxicity for soybean under saline conditions in a solution culture at P levels that were non-toxic under non-saline conditions. Large differences among cultivars were observed both in injury and in inorganic P concentrations in the leaves and roots. This growth reduction and P accumulation in the shoots under increased salinity was shown to be primarily dependent on variations in the rootstock (Grattan and Maas, 1985). Also, in solution culture Treeby and van Steveninck (1988) found P damage in lupin at elevated salinity levels which did not occur under non-saline conditions, as did Nieman and Clark (1976) for corn and Cerda et al. (1977) for sesame (*Sesamum indicum*). Champagnol (1979) considered that under increased salinity, some plants increased P concentrations (corn, sesame, and sorghum), some maintined similar levels (carrot (*Daucus carota*), cabbage, millet, and wheat), and some had decreased P concentrations (barley, onion (*Allium cepa*), and tomato (*Lycopersicon esculentum* Mill.)).

Moderate deficiencies of P reduced the yield of wheat, carrots, lettuce, and onions under non-saline conditions, but had little or no effect under saline conditions (Bernstein et al., 1974). In contrast, the yield of barley, beets, and broccoli was lowered at all salinity levels when applied P was reduced below the optimum levels determined under non-saline conditions (Bernstein et al., 1974).

Increased plant sensitivity to salinity often occurs with increases in applied P (Fine and Carson, 1954; Bernstein et al., 1974). This apparent sensitivity is likely due to limited growth at low soil P levels, such that salinity stress is not the limiting growth factor. As with the N-salinity interactions discussed earlier, it is important to establish optimum P applications under non-saline conditions.

It is not clear that the solution and sand culture experiments discussed above are directly applicable to field conditions. In sand and solution culture, optimum solution P levels (100 to 1000 μM) are orders of magnitude higher than those found in soil solution. Under field conditions, adsorbed and organically bound P are the major sources of P for root uptake, in contrast to soluble P in sand and solution culture. Khalil et al. (1967) suggested that increased P fertilization may be beneficial under field conditions because salinity decreases root growth and may decrease the root extraction of soil P.

Salinity experiments with cereals conducted in field plots at constant fertilizer application (Francois et al., 1984, 1986, 1988, 1989) demonstrate a 20 to 50% reduction in plant P content on a dry weight basis, as shown in Figure 4. Varietal differences (Aldura vs. 1000-D durum wheat), as well as variations from year to year for the same variety in the same plot (Probred semi-dwarf wheat 1982 and 1984), are also evident in Figure 4. Reductions in P levels may be related to increased Ca concentrations with increased salinity in these experiments rather than salinity per se. Despite the decreases in plant P content with increasing salinity (or Ca), it is not evident that P deficiencies caused a yield loss. As discussed above, under saline conditions some crops do not require P applications as high as those determined under optimal non-saline conditions.

Papadopoulous and Rendig (1983) found that increased salinity increased Cl and decreased P in tomato grown in soil. The plants were salinized with equal amounts of $CaCl_2$ and NaCl; thus, decreased P was associated with increased Ca in the soil solution.

Saline soils generally contain either elevated levels of Ca or alkalinity. High Ca activities or high pH results in the precipitation of added P as calcium phosphates. The short-term response of these soils to added P is attributed to the slow kinetics of precipitation. Initially, the formation of thermodynamically unstable phases allows for greater P in solution that is provided by the most stable P minerals. Increased P in solution, in turn, results in increased adsorbed P and organically complexed P. Transformation from thermodynamically unstable to stable P phases results in decreases in solution P and, thus, decreases in available P. This process explains the need for repeated P application on calcareous soils.

From a solution culture, it appears that some plants have an increased ability to absorb P under saline conditions at high P concentrations.

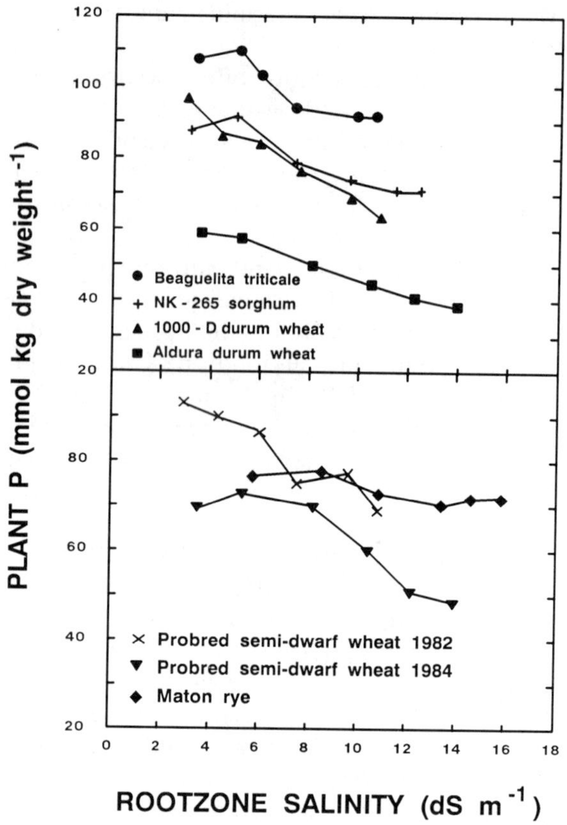

Figure 4. Plant P concentration related to rootzone salinity for sorghum (data from Francois et al., 1984), wheat (data from Francois et al., 1986), rye (data from Francois et al., 1989), and triticale (data from Francois et al., 1988). All salinity treatments for each crop received applications of P and N fertilizer.

Under field conditions, this is counteracted by salinity-induced decreases in root growth and decreased P solubility in the presence of high Ca concentration. Potential P deficiencies could likely be avoided by split application rather than by increasing quantity.

3. Micronutrients

The solubility of micronutrients is increased under saline conditions, although this has a minor effect on plant uptake. Most micronutrient uptake is related to metal-organic complexes in solution as well as easily desorbed metals in the solid phase. Plants under saline conditions usually have adequate micronutrient uptake.

More efficient use of fertilizers under saline conditions will require a better understanding of nutrient requirements under saline conditions in the field. These studies should include the effects of soil solution composition on nutrient availability.

High nitrate concentrations are typically associated with high salinity due to the high solubility of nitrate salts. In most instances, moderately saline irrigation waters can supply substantial parts of the N requirements of the crop. These inputs of N must be considered when determining fertilization levels. However, if early growth is important for a crop, it may require N fertilization at earlier vegetative stages, despite excessive N from irrigation water at subsequent stages.

F. Effects of Earthworms and Mycorrhizae

The effects of soil fauna (e.g., earthworms and termites) and flora (actinomycetes, bacteria, and fungi) on soil physical, chemical, and biological processes have been extensively studied and reviewed (Russell, 1973; Reichle, 1977; Lee, 1985). This section discusses the role of earthworms and mycorrhizae as manageable variables in increasing soil nutrient availability and fertilizer-use efficiency. The dynamics of nutrient cycling are affected more by earthworms than termites (Lee, 1983) and the role of nodulating bacteria in N fixation has been well documented (Havelka et al., 1980; Barea and Azcon-Aguilar, 1983). In addition, an increase in the population and activity of earthworms (Lee, 1985; Lavelle, 1988) and mycorrhizae (Mulligan et al., 1985; Evans and Miller, 1988; Fairchild and Miller, 1988) has been observed under reduced, compared to conventional, tillage practices, concomitant with an increase in food or energy supply. Increased plant nutrient uptake and yields have also been observed following the introduction of earthworms (Lewis, 1981; Stockdill, 1982; Hoogerkamp et al., 1983) and mycorrhizae (Cooper, 1959; Kucey, 1991) under a range of agricultural management practices. The last two observations also justified our consideration of earthworms and mycorrhizae as manageable variables in optimizing the root extraction of nutrients.

1. Earthworms

There is a need to accurately evaluate the earthworms' role in soil nutrient cycling and activity so that population activity can be maintained or stimulated. It is important to evaluate both short- and long-term roles and answer the question: when earthworm population is changed (increase, decrease, or species distribution) by man's activity, how long does it take before significant changes, if any, occur in soil structure and nutrient cycling? Although recent advances give more precise information on the impact of earthworm activities in maintaining or even improving overall soil productivity, the relative importance of their role in organic matter and nutrient cycling under different tillage practices is still poorly under-

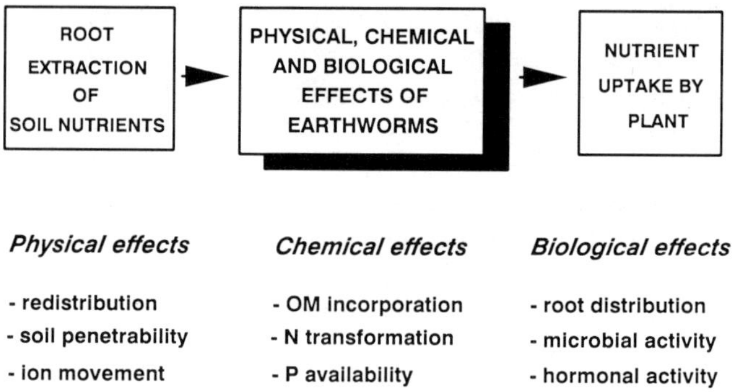

Figure 5. Effect of earthworm activity on root extraction of soil nutrients.

stood. We are, therefore, unable to answer the above question with a satisfactory degree of reliability.

Under favorable soil temperature and moisture conditions, earthworms have been found annually to consume 100% of the forest litter of an evergreen oak in Japan (8 Mg ha^{-1} yr^{-1} by *Pheretima* species) (Sugi and Tanaka, 1978) and mixed deciduous in England (3 Mg ha^{-1} yr^{-1} by *Lumbricus terrestris*) (Satchell, 1967); about 30% of the litter decomposed each year is grass savana (Lavelle, 1978) and 100% cattle dung (17 to 30 Mg ha^{-1} yr^{-1}) (Guild, 1955). During this consumption, earthworms commonly ingest 100 to 500 Mg soil ha^{-1} yr^{-1} (equivalent to 0.5 to 3.0 cm of top soil) (Russell, 1973). Consequently, earthworms can have a major influence on soil physical, chemical, and biological properties through the incorporation and assimilation of plant litter and the associated ingestion of soil material (Lee, 1985; Lavelle, 1988).

As a result of this activity, earthworms affect the root extraction of soil nutrients through an interaction of physical, chemical, and biological effects (Fig. 5).

Physical Effects: Plant residues are macerated and intimately mixed with soil during passage through the earthworm's gut. Certain earthworm species selectively feed on surface (e.g., *L. terrestris*) or partially decomposed subsurface residues (e.g., *Octolasion cyanenum*). Egestion of surface and subsurface cast material, organic matter, and nutrients can cause rapid redistribution in a soil profile. This redistribution, from the soil surface to zones where moisture is likely to be less limiting to root activity, increases the potential root extraction of nutrients. This may be of particular importance for soil additives of low-water solubility, such as lime and phosphate rock.

The burrowing activities of earthworms can allow a greater soil volume to be exploited by plant roots and a decreased susceptibility of plants to water stress. This is of particular importance in compacted soils, as

Table 3. Nutrient content of earthworm casts, soil matrix, and enrichment of casts compared to soil.[+]

Nutrient	Cast	Soil	Enrichment (%)
Total C, $g\,kg^{-1}$	52	34	1.5–2.0
Total N, $g\,kg^{-1}$	3.5	2.5	1.2–1.7
C:N	15	14	negligible
Available P, $mg\,kg^{-1}$	70	10	5–10
Nitrate N, $mg\,kg^{-1}$	22	5	3–7
Calcium, $g\,kg^{-1}$	2.8	2.0	2–3
pH	70	6.4	0.2–1.0+
Phosphatase, $mg\,P\,g^{-1}\,h^{-1}$	120	40	2–5
Urease, $mg\,P\,g^{-1}\,h^{-1}$	140	50	2–4

[+] Data averaged from Lunt and Jacobson (1944), Graff (1970), and Sharpley et al. (1979).

earthworms can penetrate zones of low permeability better than plant roots (Scott, 1977). In such soils, roots may be restricted to planes of weakness in the absence of earthworms.

Opening of earthworm burrows at the soil surface increases infiltration rates (Stockdill and Cossins, 1966; Dixon and Peterson, 1971; Sharpley et al., 1979; Edwards et al., 1989). Unless the soil is saturated, the movement of water from the soil matrix to the burrows will be limited. In well-drained soils, this movement will have little effect on water and ion movement and, consequently on the root extraction of nutrients. During heavy rainfall and when water ponds on the soil surface, preferential flow of water down the burrows can be important (Scotter and Kanchanasut, 1981; Germann et al., 1984). Under these conditions, however, there may be less leaching of nutrients than where water moves through the soils more uniformly, as nutrients present within the soil matrix between preferential flow channels are relatively isolated from leaching (Ahuja, 1986).

Root extraction of nutrients is influenced by the distance of diffusion in soil, particularly for non-mobile P, which can diffuse only short distances (<1 mm) (Nye, 1979). Thus, only P within 1 mm of a root surface is potentially available for uptake. By improving the physical environment for root growth, root distribution, and volume of soil exploited, physical limitations to the root extraction of nutrients will be reduced by earthworm activity.

Chemical Effects: Earthworm casts contain more nutrients than the source soil matrix (Table 3). The data presented in Table 3 are average values only and are dependent on litter and soil nutrient content. Although the total amounts of nutrients in soil and litter are not increased by earthworm activity, their major chemical effect is via increasing the

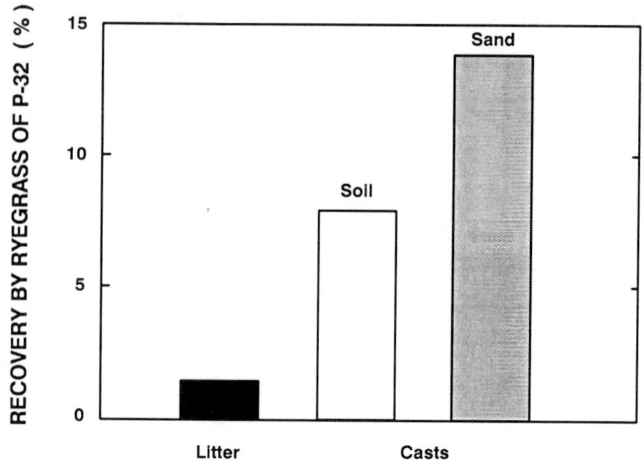

Figure 6. Recovery of ^{32}P by ryegrass grown in soil or sand amended with ^{32}P-labeled ryegrass litter or labeled earthworm casts (data adapted from Mansell et al., 1981).

rate of nutrient cycling and availability, thereby increasing the potential root extraction of nutrients at any given time.

The main source of nutrients affected by earthworms is soil organic matter, which includes plant litter in various stages of decomposition, roots, and organic matter in the soil with its complement of microflora and fauna. Ingestion, maceration, and intimate mixing of organic matter with soil increases the rate of humification and mineralization. Although most organic matter undergoes little chemical change during passage through the earthworm gut, it is finely ground, with the increased surface area exposed to microbial activity, facilitating further decomposition. The influence of earthworm activity on nutrient cycling and availability differs for N and P.

In the case of N, earthworm casts contribute small amounts available for plant growth (35 to 50 g ha^{-1} yr^{-1}) (Barley, 1959; Barley and Jennings, 1959; Syers et al., 1979). The main contribution of earthworms to N cycling is through the production of metabolic N products. The excretion of nitrogenous waste products of nutrition and secretion of mucoproteins onto the body surface for lubrication and respiratory exchange by the earthworm can input 18 to 50 kg N ha^{-1} yr^{-1} (Lee, 1983). On the assumption that annual earthworm mortality and biomass production are similar in a stable population and two to seven times the mean annual biomass, Lee (1983) calculated N contribution from earthworm mortality to range from 100 to 225 kg N ha^{-1} yr^{-1} in temperate pastures. As rapid recycling of N through earthworms and microbial tissue occurs many times annually, the above amounts indicate the importance of earthworm activity on root extraction of N.

The five- to ten-fold increase in P content and availability in earthworm casts (Table 3) results from the enhanced mineralization of organic P, the enrichment of clay-sized particles in casts, and a reduction in P sorption capacity of soil by organic matter blockage (Sharpley and Syers, 1977). Thus, most of the additional P present in casts is held in more physically sorbed rather than chemisorbed forms, which are readily available to plants. As with N, the major effect of earthworm activity on P availability results from an increase in the rate of P cycling. This effect is likely to be of greater importance in soils of low P status receiving little or no fertilizer. This is illustrated by Mansell et al. (1981), who labeled surface casts of *Allolobophora caliginosa* and *Lumbricus rubellus* with ^{32}P by supplying ^{32}P-labeled ryegrass (*Lolium perenne*) as their sole food source. Passage of ^{32}P-labeled ryegrass through the earthworm increased the short-term plant availability of P by a factor of approximately three (Fig. 6). Further, recovery of ^{32}P by ryegrass was significantly greater when labeled casts were placed on the surface of sand with low available P ($1\,mg\,kg^{-1}$ Olsen P, Olsen et al., 1954) than on soil with a higher available P content ($20\,mg\,kg^{-1}$) (Fig. 6).

Biological Effects: The volume of root material penetrating the soil has been shown to be increased by earthworm activity (Stockdill and Cossins, 1966; Edwards and Lofty, 1978). In fact, the zone of maximum root growth has been shown to coincide with the zone of major activity of the dominant earthworms present. Surface-active earthworm species promoted barley (*Hordeum vulgare*) root growth in surface soil horizons, whereas deeper burrowing species promoted deeper rooting (Edwards and Lofty, 1978). Subsequently, Edwards and Lofty (1980) demonstrated that cereal roots followed channels in the soil, whether made artificially or by the activity of earthworms. Clearly, root distribution of both shallow and deep rooting plants is affected by earthworm activity through a complex interaction of physical and chemical factors discussed above.

Both microbial population and activity are increased during passage through the earthworms' gut as a function of the organic matter content of the initial food source (Gorbenko et al., 1986; Scheu, 1987; Tiwari et al., 1989). Parle (1963) observed that numbers of bacteria and actinomycetes increased 1000-fold during passage through the gut, and oxygen consumption remained higher in earthworm casts than in soil for 50 days, indicating increased microbial activity. This enhanced activity is probably responsible for the increased phosphatase and urease activity found in earthworm casts compared to the underlying soil (Table 3). Furthermore, the stimulation of microbial activity results in the rapid transformation of soluble N into microbial protein and, consequently, prevents its rapid loss to lower soil horizons by leaching.

The presence of plant growth promoting substances, such as indole or auxin-like compounds, in earthworm casts has also been recognized (Nielson, 1965; Springett and Syers, 1979; Nardi et al., 1988). In a study of ryegrass root growth in the presence of *A. caliginosa* and *L. rubellus* casts, Springett and Syers (1979) observed the roots to be negatively geotropic by growing laterally and upwards into the casts of *L. rubellus* but not into the *A. caliginosa* casts. As the nutrient content of the soil matrix was amended to that of the casts, Springett and Syers (1979) concluded that *L. rubellus* casts probably contained an auxin-like substance or some substance that modifies the effects of plant auxins.

Two current issues facing agriculture are soil physical limitations to crop growth and fertilizer-use efficiency. From the above overview of earthworm effects on the root extraction of nutrients, it is clear that earthworms can have significant positive effects on both these aspects. There is still a need for research, however, on the mechanisms by which earthworms bring about improvements. This has resulted, in part, from a general difficulty in the quantification of earthworm effects on the root extraction of nutrients, due to measuring total cast production (subsurface as well as surface) and wide spatial and temporal variations in earthworm population dynamics, litter quantity and quality, and soil nutrient status. This information is required, however, before suitable management practices can be developed to optimize the long- and short-term physical, chemical, and biological effects of earthworms on the root extraction of nutrients. For example, the role of earthworms in incorporating and redistributing plant litter and surface applied nutrients in the soil rooting zone will be of particular importance in low input agricultural systems. It is suggested that the enhancement of earthworm activity may reduce the number of tillage operations necessary.

Further, it is necessary to determine the feasibility and cost/benefit ratio of earthworm introduction under given management practices by considering native soil fertility, indigenous earthworm populations, and climatic and crop factors. A lack of specific information on the interaction among earthworm species, soil properties, and crop requirements exists, which could lead to the inefficient use of increasingly expensive fertilizers. Because the relationship between physical, chemical, and biological effects of earthworm activity on the root extraction of nutrients is complex, interactive, and cyclical, future progress in reducing limitations to root growth will depend on interdisciplinary research.

2. Mycorrhizal Associations

Indigenous and introduced mycorrhizal species can modify the potential root extraction of nutrients. The effect of mycorrhizae on increasing plant growth and the root extraction of nutrients, particularly P, is well documented for many plant species (Moose, 1973; Tinker, 1978; Gianinazzi-

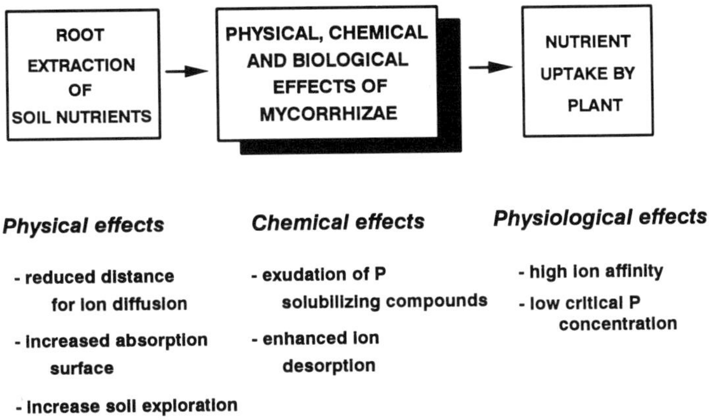

Figure 7. Effect of mycorrhizae activity on root extraction of soil nutrients.

Pearson, 1986; Kucey et al., 1989). Mycorrhizae are widespread under natural conditions, with vesicular-arbuscular mycorrhizae (VAM) most common in agricultural soils. Three primary mechanisms by which VAM enhance soil P utilization by plants are the increased physical exploration of the soil, chemical modification of the rhizosphere, and physiological differences between VAM and plant roots (Fig. 7).

Ellis and co-workers (Ellis et al., 1985, 1988; Ellis, 1988, 1990; Watson and Ellis, 1990) found that VAM essentially act as extensions of the plant root system, enabling the plant to more thoroughly utilize not only P but also all other nutrients and molecules present in small-diameter soil pores. For example, they found the VAM-infected wheat could dry a soil more thoroughly than non-infected plants, enabling the wheat to better withstand intermittent drought. They also found less nitrate-N leaching through the soil with VAM-infected plants vs. non-infected plants. Likewise, residual atrazine was absorbed by infected plants. They concluded that these phenomena occurred because the small diameter VAM fungal hyphae could penetrate soil pores less than 30 μm in diameter, which largely exclude plant root hairs. They also found that practices which enhance organic matter conservation (e.g., reduced tillage, crop rotations, cover crops) also enhance VAM infection.

Physical Effects: Extensive hyphal growth of VAM reduced the distance for diffusion of P in soil, thereby increasing uptake. This effect is greater when diffusion limits uptake (Gerdemann, 1968). Consequently, a greater response to VAM infection has been exhibited in coarse rather than fine rooted plant species (Crush, 1973), in high rather than low P sorbing soils (Yost and Fox, 1979), and in soils rather than solution cultures (Howeler et al., 1982). In addition, the generally smaller diameter of VAM hyphae

(2 to 4 μm) compared to root hairs (710 μm) affords a greater absorptive surface area for hyphae and enables the entry of hyphae into soil pores and organic matter that cannot be entered by root hairs.

Chemical Effects: Vesicular-arbuscular mycorrhizae may chemically modify the rhizosphere through the exudation of chelating compounds (Lapeyrie, 1988; Jayachandran et al., 1989; Treeby et al., 1989) or phosphatases (Harley, 1989) which could solubilize poorly soluble soil P. It is clear, however, that VAM utilize the same sources of P as non-mycorrhizal plants (Mosse et al., 1973; Kucey et al., 1989; Blal et al., 1990), but do so more efficiently. Depletion of soil solution P in the rhizosphere at a faster rate or to a lower concentration in mycorrhizal compared to nonmycorrhizal plants may induce soil P desorption to restore equilibrium conditions.

Differential absorption of anions and cations by mycorrhizal and nonmycorrhizal plants may induce changes in the rhizosphere pH (Hedley et al., 1982; Buwalda et al., 1983). This will be important in acid soils in which P solubility is controlled by ion exchange equilibria involving negatively charged clay minerals, exchangeable cations and organic matter, and in neutral to alkaline soils where P solubility is dominated by calcium phosphates. In addition, more efficient utilization of ammonium by mycorrhizal rather than nonmycorrhizal plants (Smith et al., 1985) will lead to increased H^+ extrusion to the rhizosphere as a result of ammonia assimilation in mycorrhizal plants (Bolan et al., 1989).

Physiological Effects: On a unit weight basis, mycorrhizal plants can rapidly absorb larger amounts of P than nonmycorrhizal plants (Bolan et al., 1987). This difference cannot be accounted for by increased surface area of the hyphae alone; it is, therefore, possible that a greater affinity for P in mycorrhizal plants increased the absorption rates. It is also possible that the critical or minimum P concentration, below which there is limited net absorption of P, is lower for mycorrhizal than non-mycorrhizal plants. Although further research is needed in this area, several studies have suggested that a difference in critical P concentration may exist (Mosse, 1973; Bolan et al., 1983; Guttay and Dandurand, 1989).

With the above outline in mind, there are two approaches to the utilization of VAM for the benefit of long-term soil management on the potential root extraction of nutrients (Hedley et al., 1989a). The first and most practical approach involves the efficient use of indigenous soil mycorrhizae. This efficiency can be increased by such management practices as using selective fungicides not detrimental to VAM, using crops responding to indigenous VAM, and maintaining soil P below critical concentrations. The second approach involves the isolation, mass production, and inoculation of plants with VAM. Recent developments in methodology to cultivate VAM fungi from roots infected with tumor inducing plasmids (Mugnier et al., 1986) have the potential to mass produce VAM fungi annexically. Even so, the probability of inoculation strains forming successful associations with the roots in soils with individual mycorrhizae may be low. The ability of VAM isolates to increase the potential root extraction of nutrients and compete with indigenous VAM varies with isolate strains. The use of VAM to supplement fertilizer P application will, thus, depend on the selection of strains that increase P uptake and successfully compete with indigenous microflora.

Isolated strains of *Penicillium bilaii* from western Canadian soils have been shown to efficiently solubilize inorganic P compounds from soil and fertilizer sources as well as in liquid cultures (Kucey, 1987; Asea et al., 1988). Although other soil fungi, bacteria, and actinomycetes have been reported to increase the solubilization and plant uptake of P (Banik and Dey, 1982; Kucey et al., 1989). *P. bilaii* has been sucessfully used to enhance the root extraction of nutrients in field trials (Kucey and Leggett, 1989; Kucey, 1992). For example, the yields of wheat inoculated with *P. bilaii* were compared with uninoculated wheat yields in 38 field trials in three Canadian prairie provinces encompassing a wide range of soil and climatic conditions over a two-year period (Kucey, 1992). *Penicillium bilaii* inoculation increased the wheat yields on low to medium residual P soils ($<20\,kg\,P\,ha^{-1}$, 28 trials), receiving 0, 4, and $9\,kg\,P\,ha^{-1}\,yr^{-1}$ fertilizer compared to uninoculated plots at 0.4, 4.0, and 8.0% levels of significance, respectively (Fig. 8). No yield response was observed on high residual P soils ($>20\,kg\,P\,ha^{-1}$, 10 trials) to either *P. bilaii* inoculation or fertilizer P application (Fig. 8). The greatest yield increases were obtained when no fertilizer was applied. Kucey (1992) subsequently concluded that *P. bilaii* could both increase indigenous soil P and fertilizer P availability, thereby reducing application rates in some cases and contributing to the development of more sustainable agricultural production systems. As a result, the commercial use of *P. bilaii* as a soil inoculant to increase the root extraction of P and fertilizer P use efficiency is being developed by a Canadian agribiotechnology company.

It is clear that VAM should be considered an essential factor in optimizing fertilizer efficiency, not only of soluble P fertilizers but also of rock phosphates, for a wide range of soils. Although detailed information on the physical, chemical, and biological effects of various VAM strains has been obtained using solution cultures and sterilized soil, the following field research is needed to better understand and utilize VAM to increase root extraction of nutrients:

a. Investigate the soil physical and chemical factors affecting VAM activity in non-sterilized agricultural soils under field conditions.
b. Investigate the residual effects of VAM innoculation.
c. Quantify VAM activity under reduced tillage practices.

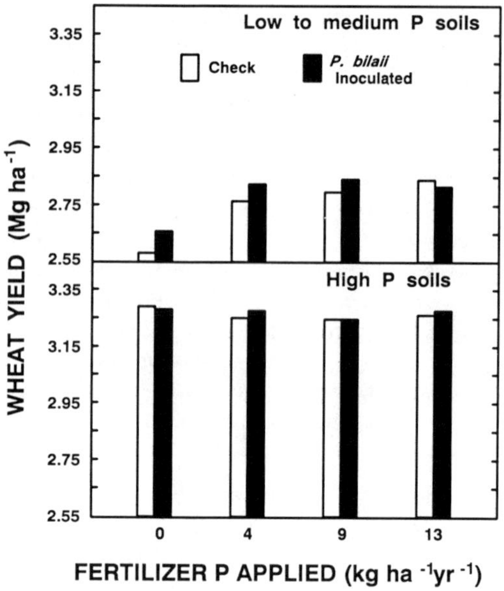

Figure 8. Effect of *Penicillium bilaii* on wheat yields under field conditions on low to medium P ($<20\,kg\,ha^{-1}$) and high P ($>20\,kg\,ha^{-1}$) soils. (Data supplied by D. Polonenko, Philom Bios Inc., Saskatchewan, Canada).

d. Identify the management practices that affect VAM enhancement of P uptake. Because the relative benefit of VAM inoculation increases as soil P availability decreases, lower rates of fertilizer P use may be afforded for certain soils, crops, and management.

Information from the above research needs will facilitate the inclusion of VAM as an integral part of sustainable agricultural production systems.

G. Effects of Soil Conservation

1. Reduced Tillage Measures

Reduced and no-till methods of cultivation are commonly used as a cost reduction and first-line defense against accelerated soil erosion. Crop residues on the soil surface have long been recognized as effective means of reducing runoff and soil losses. A popular form of reduced tillage is ridge-till, in which row crops are produced on a ridge with a residue cover between the rows. These practices affect plant root activity and nutrient uptake by conserving soil water, thereby enhancing the microbial turnover of nutrients as well as plant root growth. Also, the reduction in soil erosion losses maintains long-term soil productivity and nutrient supply.

Zuzel et al. (1990) showed that by maintaining a high level of soil fertility through fertilization, more crop residues were produced, providing better surface cover and thereby improving water infiltration and yield of the next crop.

One of the main problems in optimizing the root extraction of nutrients under conservation tillage is maintaining adequate soil available P levels in the actively growing plant root zone and avoiding surface accumulations of residual fertilizer and crop residue P (Carter and Rennie, 1982; Eckert and Johnson, 1985). Even so, the maintenance uptake and grain yield by no-till corn (Triplett and Van Doren, 1969; Moschler et al., 1972; Fink and Wesley, 1974) has been attributed to increased soil moisture and, thereby, greater root growth immediately below the crop residue in soil where P has accumulated. Thus, it is clear that field research on the effect of fertilizer placement below the seed during planting is needed to evaluate the yield response and root extraction of P for crops of differing root distributions (Stecker et al., 1988; Havlin, 1992). The selection of crops which can more efficiently utilize residual soil inorganic and organic P pools, while being economically viable to a farmer, may also enhance the agronomic effectiveness of conservation tillage systems.

2. Conventional Measures

Conventional measures include such practices as terracing, contour-strip cropping, and grassed waterways, among others. Again, these practices reduce runoff and soil loss, thereby increasing soil water storage and maintaining soil productivity. The additional stored water usually offers the potential for increased crop production, so there may be an increased need for plant nutrients. However, fertilizer requirements may not be increased because the accompanying reduction in soil erosion results in better maintenance of the original indigenous soil fertility level. If extensive removal of topsoil was involved in the construction of these structures, there may be need for additional fertilizers in the areas disturbed.

3. Wind Erosion Control Measures

The first practice usually utilized for wind erosion control is the adoption of reduced and no-till practices that leave crop residues on the soil surface. Because this practice also usually enhances microbial activity near the soil surface, soil aggregation is also usually improved (Gupta and Germida, 1988). Aggregates over 0.84 mm diameter are generally resistant to wind erosion. The effects of these practices on plant root activity and nutrient uptake were discussed in earlier sections. Other wind erosion control practices include strip cropping and the use of windbreaks, shelter belts, and grass barriers. In general, these practices have little direct influence on soil fertility. However, if these practices are effective, they will also control soil erosion losses, thereby maintaining

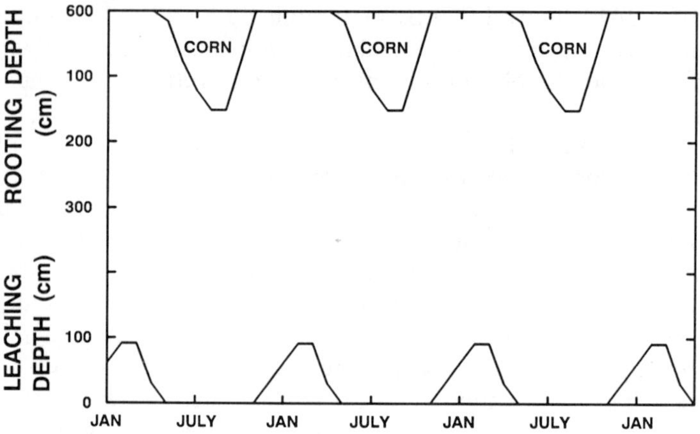

Figure 9. Root development of continuous corn in relation to typical soil drainage events over three years.

soil fertility and productivity. Tanaka and Aase (1989) showed that the productivity of soils from which topsoil had been removed could not be completely restored by the use of more fertilizers.

H. Effects of Cropping Systems Management

Cropping system management is one of the most direct ways to affect the root extraction of nutrients through improved depth of rooting and/or intensity of rooting. Crops vary considerably in the time, depth, and intensity of root development. Species selection should be given careful consideration in both long-term (e.g., grain-forage rotation systems) and short-term cropping systems (e.g., multiple cropping systems).

1. Crop Selection for Long-Term Systems

The depth and extent of nutrient extraction is directly related to the depth and extent of crop root systems. There are some valid generalizations concerning the differences between rooting patterns of various crops, despite the fact that individual rooting patterns can be affected by available soil water, dense soil layers, and soil chemical properties. Cereal crops, such as corn and wheat, have intense fibrous rooting to a depth of 1 to 2 m, while forage legumes, such as alfalfa and red clover, have deep taproot characteristics which allow rooting to 2 to 3 m in the first year and to even greater depths (up to 5 to 6 m) in later years (Weaver, 1926; Kohl and Kolar, 1976; Black et al., 1981; Dwyer et al., 1988).

One method to increase the root uptake of nutrients is to alternate crops with different rooting patterns, particularly different depths of

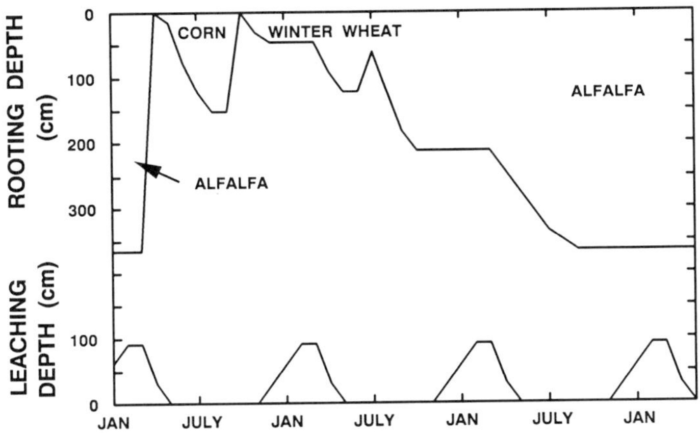

Figure 10. Root development of corn-winter wheat-alfalfa system in relation to typical soil drainage events over three years.

rooting. The most difficult nutrient to retain is N because (1) it is required in large quantities and over short periods by cereals and (2) it is a mobile nutrient subject to loss in percolating water. In order to quickly evaluate the impact of different crops on N retention, we have plotted the hypothetical root development for various crops vs. time and have overlayed this onto a typical nitrate leaching pattern. A three-year period for a continuous corn system (Fig. 9) shows good potential extraction of nutrients during the year, but root development does not coincide with N drainage losses. An alternative cropping system made up of corn-winter wheat-alfalfa (Fig. 10) gives a much different root development pattern, one which should be more efficient at retaining N since it is deeper and has roots present most of the year. Similarly, rotating a shallow rooting crop, like potatoes, with a deeper rooting crop, like corn, should improve root exploration and nutrient recovery (Singh and Sekhon, 1977).

The practice of summer fallowing in alternate years is common in dry land agriculture. This practice maintains an active root zone only 50% of the time and could result in nitrate losses once the soil water is recharged during the fallow period (Campbell et al., 1984). Black et al. (1981) studied three cropping systems: (1) alternate year fallow; (2) spring wheat-winter wheat-fallow (fallow every three years); and (3) a small grain-safflower system without fallow. Black and colleagues reported precipitation use efficiencies of 40, 59, and 82%, respectively, for these three systems. Thus, the root extraction of water (and presumably of nutrients) was highest in the system with the most continuous active root systems (annual cropping), which included the deep-rooting safflower crop. Another example of the benefits of deep-rooted crops was the use of alfalfa to extract nitrate below abandoned feedlots (Schuman and

Elliott, 1978). Muir et al. (1976) also reported that nitrate concentrations under alfalfa were not different than those under native range. However, precautionary notes also need to be sounded regarding the use of deep-rooted legumes: (1) the need to allow the water depleted soil profile to recharge before the next crop (Hoyt and Leitch, 1983; Grecu et al., 1988); (2) the need to recrop after alfalfa killing (Radke et al., 1988); and (3) the need to lower fertilizer N applications to the next crop due to increased N mineralization (Kurtz et al., 1984). The use of deep-rooted legumes, as related to nitrate pollution, is also discussed by Russelle and Hargrove (1989). In general, anytime a deeply rooted crop can be included in a cropping system, it should improve the root extraction of nutrients. However, more research is needed to quantify the long-term benefits of this practice.

The above benefits of deeply rooted legumes are difficult to achieve in practice because of the decline in forage crop acres and the rise in cash grain acres. The decline of the medium-sized mixed livestock farm has simultaneously reduced cropping system flexibility and the number of economically viable options for inserting deep-rooted forages into rotations. Possible solutions to this problem include (1) developing alternative uses for forages (e.g., alfalfa for paper pulp), (2) introducing new economically viable crops with deep root systems (e.g., safflower [*Carthamus tirctorius*] and kenaf [*Hibiscus cannabinus* L.]), or (3) breeding present crops to increase the depth and efficiencies of their root systems (O'Toole and Bland, 1987).

The use of crop selection to improve the root extraction of nutrients has also been mentioned in other sections of this discussion. These include selection of acid tolerant crops or cultivars for acid subsoil situations, selection of crops with high root growth pressures to penetrate hardpans, and selection of salt tolerant crops to improve nutrient use in salt affected soils.

2. Crop Selection for Short-Term Systems

Selecting crops to improve nutrient retention in short-term systems (less than 1 yr) involves application of the general principles of (1) maintaining an actively growing crop at all times, (2) increasing the intensity of root development, and (3) growing crop mixtures. Ideally, these crops will have complementary root systems to fully exploit the available nutrients. Examples of short-term systems include winter cover crops, multiple cropping, mixed cropping, and crops for diverted acres.

Winter cover crops are particularly useful where winter precipitation exceeds evapotranspiration, which in the United States is basically east of the Missouri River. Cover crops may also be useful in other regions where irrigated agriculture predominates. The value of a winter cover crop can be seen by contrasting Figures 11 and 9, which show root

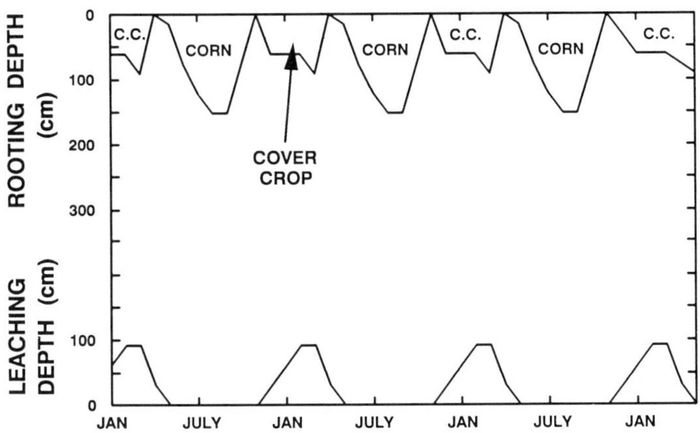

Figure 11. Root development of continuous corn with a winter cover crop (C.C.) in relation to typical soil drainage events over three years.

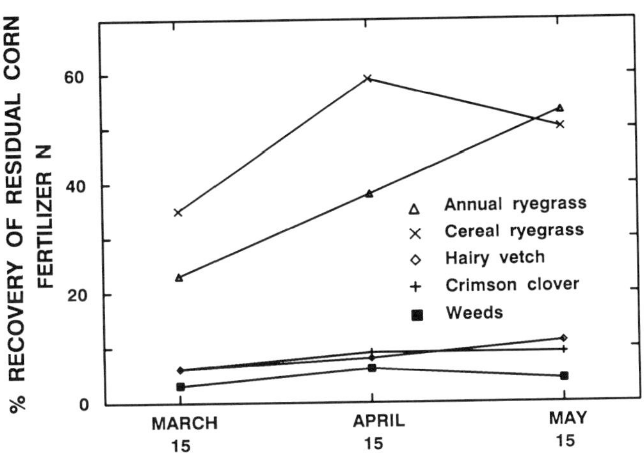

Figure 12. Average percent recovery of residual corn fertilizer N by winter cover crops of cereal rye, annual ryegrass, hairy vetch, crimson clover, and weeds, using ^{15}N in Maryland (data adapted from Meisinger et al., 1990).

development in continuous corn systems with and without a cover crop. The winter cover crop contributes an active root system during the winter leaching season and, thus, offers the opportunity to conserve N. Recent research by Meisinger et al. (1990) and Shipley et al. (1992) assessed the ability of various winter cover crops to retain residual corn fertilizer N. The study added ^{15}N-labeled fertilizer to corn and then traced the uptake of the residual tagged N by the succeeding winter cover crops. The results (Fig. 12) showed that grass cover crops (particularly cereal rye) were clearly superior to legumes in utilizing residual corn fertilizer N. The legumes were inferior to grasses due to their shallow and slowly developing root systems which did not achieve full N uptake potential until the next spring. By that time, most of the labeled N had leached out of the legume root zone. Another example of conserving nutrients with cover crops was reported by Bertilsson (1988) in Sweden, who found that winter rape (*Brassica napus*) reduced nitrate losses even after fall manure applications. The effect of management of winter cover crops on the next crop should also be considered. This may require timely killing to allow water replenishment to insure germination of the next crop (Wagger and Mengel, 1988).

Multiple cropping, i.e., growing more than one crop per year, is another way to extend the period of active root uptake. Multiple cropping is most successful if the two crops have non-competing root systems and if the water and nutrients that would otherwise be lost are utilized by the second crop (Hook and Gascho, 1988). The use of double cropped soybean after winter wheat in the southeastern United States is a good example of multiple cropping. If the two crops have competing root systems, then the second crop should be drought tolerant, for example, the use of sorghum after winter wheat. Another example of the efficiency of complementary root systems is given by Hansson and Andren (1987), who observed greater rooting depths under a barley/alfalfa system than under a barley/meadow fescue. There are little or no data concerning long-term nutrient efficiencies in multiple crop systems. This remains a primary research need in areas with ample water and a growing season to support such cropping systems.

Managing crop selection to increase rooting intensities for annual crops can be done by selecting crops or cultivars adapted to specific soil or climatic conditions, such as acid tolerance, drought tolerance, or salt tolerance. Planting these crops in more dense or uniform stands should also aid in greater interception of solar radiation, greater dry matter production, and greater utilization of water and nutrients (Sojka et al., 1988). For example, Karlen and Camp (1985) found greater N utilization for a narrow twin-row configuration compared to conventional rows.

The use of mixed cropping, i.e., growing more than one crop at the same time, can also increase nutrient extraction if the crops have different root systems and different peak dry matter production seasons. The interseeding of grasses and legumes for forage production are the most common examples (e.g., alfalfa-grass mixtures and crimson clover [*Trifolium incarnatum*] seeded into warm season grass pastures). The increased total dry matter production and increased rooting depths of these systems lead to greater root extraction of nutrients. The results and concepts for mixed cropping are discussed elsewhere (Haynes, 1980; Russelle and Hargrove, 1989).

Crop selection for short-term systems is also important in set-aside acres. Schepers (1988) summarized data showing that 170 kg N ha^{-1} was mineralized during the summer on a non-cropped area in Nebraska. However, a summer cover crop of oats was effective in taking up N mineralized from the surface 60 cm of soil.

The above discussion and examples have clearly shown that crop root development should be considered when developing cropping systems. Crops with complementary rooting systems should follow one another whenever possible. Efforts should be made to keep an actively growing crop on the soil at all times in order to increase the root extraction of nutrients, especially N.

III. Interaction Among Management Factors

The preceding sections have focused on the main effects of different manageable variables and approaches to increase the root extraction of nutrients. However, when implementing these approaches, one must consider the nature of the problem restricting root growth (subsoil acidity, compaction, etc.), the impact of the reduced root growth on the crop (reduced water supply, reduced nutrition, poor aeration, etc.), and the possible remedies for a given soil-crop system (irrigation, crop or cultivar selection, tillage, etc.). Problems seldom lend themselves to clear-cut simple solutions. Rather, a solution usually involves the interaction of several factors to achieve the final goal.

An example of these interactions is demonstrated in a study reported by Elkins et al. (1977) in Alabama on a Cahaba loamy (siliceous, thermic Typic Paleudult) sand with a plowpan that confined cotton roots to the surface 23 cm of soil. The restricted rooting and high summer ET in Alabama translated into only a 4.5 day supply of water for the cotton crop. Deep tillage to break up the plowpan was only marginally successful; because its effects were inconsistent from year to year, it was costly and the yield increase amounted to only 30% of the control treatment. Elkins et al. (1977) considered the approach of selecting a specific crop to precede the cotton that could break through the plowpan. Their choice was 'Pensacola' bahiagrass (*Paspalum notatum* Flugge), which had previously been observed to penetrate artificial traffic pans. The four-year average yields of cotton planted after bahiagrass were twice those for continuous cotton. Root measurements showed that continuous cotton roots were restricted to the surface 23 cm, but that cotton roots after bahiagrass penetrated to 60 cm, with some root penetration to 180 cm. The greater rooting depth of cotton decreased water stress. The only chemical or physical change in the soil to explain this greater rooting was a change in macroporosity, where pores greater than 1 mm appeared eight times more numerous after bahiagrass than corn. The bahiagrass

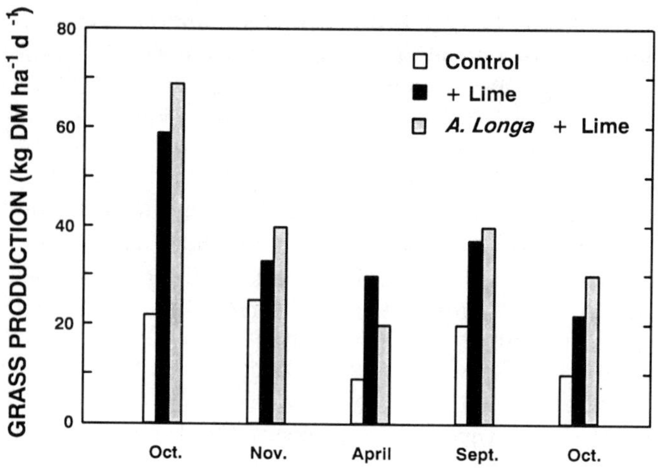

Figure 13. Grass production as influenced by earthworm (*A. longa*) and surface application of lime ($5\,\mathrm{Mg\,ha^{-1}}$) (unpubl. data, with permission from J.A. Springett).

Table 4. Soil pH values at different depths in Tokomaru silt loam 10 months after surface-applying *A. longa* and lime ($5\,\mathrm{Mg\,ha^{-1}}$) (data adapted from Springett, 1983).

Depth (cm)	*A. Longa*[+] resident	Resident earthworms	Control (no lime)
0–2	6.8	7.2	5.7
2–4	6.5	6.4	5.5
4–6	6.4	5.6	5.3
6–8	6.0	5.4	5.3
8–10	5.6	5.4	5.1
10–12	5.6	5.4	5.2
12–16	5.5	5.2	5.2

[+] *A. caliginosa* and *L. rubellus*.

was successful in penetrating the plowpan because it is an acid tolerant crop and its roots have a fibrous sheath beneath the epidermis that increases root rigidity (Peterson et al., 1976). This study contains interactions between crop selection, soil compaction, acidity tolerance, and water stress, which all contribute toward developing a solution to increase root growth and crop performance.

The effect of earthworm activity on the root extraction of nutrients following the surface application of lime and fertilizers also indicates the complex interaction among manageable variables and potential benefit to crop production. Research has shown that when earthworms mix surface applied chemicals of low water solubility into the soil, plant

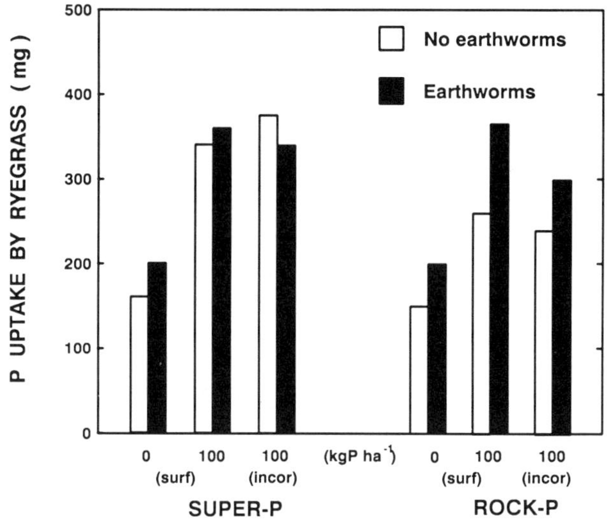

Figure 14. Uptake of P by ryegrass as influenced by earthworms and the source, method, and rate of P application (Mackay et al., 1982. Reprinted with permission from *Soil Biology and Biochemistry*, vol. 14, Plant availability of phosphate in superphosphate and a phosphate rock as influenced by earthworms, copyright 1982, Pergamon Press PLC).

growth can be increased. Introduction of the earthworm *A. longa* to a grassed field increased the incorporation of surface applied lime compared to resident species (*A. caliginosa* and *L. rubellus*) (Springett, 1983). Further, *A. longa* mixed the lime to a much greater depth (Table 4). As a result, grass production was increased by the introduction of *A. longa* compared to the application of lime only (Fig. 13). This effect was significant during spring (Southern Hemisphere) when increasing soil moisture and temperature enhanced earthworm activity.

In a glasshouse study with perennial ryegrass, Mackay et al. (1982) showed that earthworm activity (*A. caliginosa* and *L. rubellus*) increased the yield and uptake of P from a pelleted rock P (Chatham Rise Phosphorite, CRP) but not from the more soluble single super P (Fig. 14). Further, P uptake in the presence of earthworms was greater when CRP was surface applied as compared to incorporated in the top 2 cm of soil. The increased agronomic effectiveness of CRP resulted from an increase in the degree of intimate contact of the CRP with the soil surfaces during passage through the earthworm. Earthworm burrows open to the surface also facilitated the downward movement of CRP from the site of application during rainfall. In addition, the encouragement of earthworm activity in orchards through mulching has reduced the number of tillage operations necessary (Tisdall, 1978). These examples indicate the complex agronomic effectiveness of earthworm activity on the root extraction of nutrients.

IV. Modeling Long-Term Management Effects on Root Extraction of Nutrients

Several models are available for simulating soil nutrient cycling and plant root extraction for both practical agricultural management and identification of basic research needs. A user can choose one suited to a certain problem or need, but herein lies one of the main problems associated with models: the selection of the most appropriate model to obtain the level of detail required. Once an appropriate model is chosen, the major limitation is obtaining the necessary data for input parameters to run the model. This most frequently limits model use. Not only do many models require detailed information on soil physical, chemical, and biological properties as well as crop and tillage operation, but the output will only be as reliable as the data input. In addition, a lack of field validation data often limits the assessment of model accuracy. Consequently, use of these models to provide quantitative estimates under specific environmental conditions is limited. Most models are recommended only for *relative* comparisons of management effects on soil productivity.

Considerable information is available on the complex nutrient transformations, diffusion and mass flow of nutrients, and root extraction of nutrients as a function of soil, climatic, and plant variables. For example, several models have been developed to simulate specific soil processes such as ammonia volatilization (Parton et al., 1983), denitrification (Smith, 1981), P flux (Claassen and Barber, 1974), inorganic (Mansell et al., 1977) and organic P transformations (Mishra et al., 1979), and P availability (Bennett and Ozanne, 1972; Barrow and Carter, 1978; Cox et al., 1981). However, attempts to formulate holistic management oriented simulation models have been limited. A lack of information about and the ability to predict processes such as root growth and root morphology has slowed the development of comprehensive models. Description of the complex interaction between agricultural management and root extraction of nutrients over time and space is, thus, difficult. The types of models range from complex mathematical representations of soil nutrient pools and plant uptake to relatively simple predictions of fertilizer requirements for farm management decisions.

A. Nutrient Uptake

Mathematical models simulating nutrient flux to plant roots have been developed (Olsen and Kemper, 1967; Helyar and Munns, 1975; Claassen and Barber, 1976; Cole et al., 1977; Schenk and Barber, 1979a;

Cushman, 1982). In general, these models follow the work of Nye and Marriott (1969) by describing nutrient movement to individual roots by mass flow and diffusion:

$$J_r = D_e \frac{\delta C_s}{\delta_r} + v_o C_L \tag{1}$$

where J_r is flux to the root, D_e is the effective diffusion coefficient, r is the radial distance, C_s is the concentration of ions in the solid phase that readily equilibrates with the concentration of ions in the soil solution (C_L), and V_o is the rate of water flux into the root. Nutrient uptake by the root follows Michaelis-Menten kinetics, after subtracting a term for efflux, E, as described by Claassen and Barber (1974):

$$J_r = \frac{I_{max} C_L}{K_m + C_L} - E \tag{2}$$

where I_{max} is the rate of influx at infinite concentration and K_m is the Michaelis-Menten constant. Required soil parameters include the effective average diffusion coefficient, initial ion concentration in solution, and buffer power. Plant parameters include the maximum influx rate, Michaelis-Menten constant, water influx, root radius, initial root length, and rate of root growth.

Claassen and Barber (1976) proposed a model based on the theoretical consideration of the processes of potassium uptake by plant roots growing in soil. This model was subsequently used to predict P uptake by maize (Schenk and Barber, 1979a) and soybean (Silberbush and Barber, 1983) and developed to account for nonlinear root boundary conditions; arbitrary spatial and temporal changes in the buffer power and diffusion coefficients; nonlinear sources and sinks of nutrients in both the rhizosphere and bulk soil; and arbitrary, nonlinear, initial conditions (Cushman, 1984). Although the model does not simulate the cycling of nutrients in inorganic (solution and adsorbed) and organic (fresh and stable) pools, it has provided a quantitative assessment of the importance of several factors, such as root morphology (Schenk and Barber, 1979a,b), distribution (Kuchenbuch and Barber, 1987), soil temperature and water content (Mackay and Barber, 1984; Barber et al., 1988), and fertilizer placement (Anghinoni and Barber, 1980; Kovar and Barber, 1987) on the root extraction of nutrients by several crops.

B. Nutrient Cycling

1. Research Oriented

A more comprehensive model of the cycling of P forms in semi-arid grasslands was developed by Cole et al. (1977). Root extraction of P is assumed to be limited by diffusion. As plant uptake reduces the P concentration at the root surface, a concentration gradient is established in the surrounding water films and P ions move toward the root. The solution P is replenished from a labile P pool, which in turn is replenished from slightly soluble P minerals, adsorbed P, and organic P mineralization. As the model is intended to run on one day or longer time steps, P is taken up directly from the labile pool. Rate of uptake per unit weight of live root in a given soil layer are controlled by the values of solution P, labile P, and soil temperature and water content.

Recent development of the Century model (Parton et al., 1987, 1988), simulating the dynamics of C, N, P, and S in the soil plant system using a monthly time step, incorporated many of the concepts used in earlier P (Cole et al., 1977) and soil organic matter models (Parton et al., 1983). The Century model has been used to simulate the impact of cultivation (100 years) on soil organic matter dynamics, nutrient mineralization, and plant production (Parton et al., 1988).

These long time-frame models (Cole et al., 1977; Parton et al., 1988) incorporate the effects of moisture, temperature, soil properties, plant phenology, and organic matter decomposition on nutrient flows and simulate many soil and plant processes. Their use is limited to the availability of detailed soil and plant information, which is not generally available. However, they have revealed gaps in our knowledge of processes such as the contribution of organic matter cycling to long-term soil productivity as a function of management (Parton et al., 1988). Consequently, these models provide valuable direction for future research.

2. Simplified

Long-term soil management can influence the root extraction of nutrients through fertilizer application, crop removal, leaching, and changes in the organic matter content of the soil. Consequently, a simple model simulating soil P, N, and C cycling was developed for incorporation into the Erosion-Productivity Impact Calculator (EPIC), which is composed of physically based components for simulating erosion, plant growth and related processes, and economic components for such assessments as the cost of erosion and determining optimum management strategies (Williams, 1983; Williams et al., 1984). For P, the model simulates uptake and transformations between several inorganic and organic pools in up to 10 soil layers of variable thickness (Jones et al., 1984). Fertilizer P is added to the labile inorganic P pool which rapidly achieves equilibrium with active inorganic P. Movement of inorganic P from the active to the stable pools simulates the slow adsorption of inorganic P. Crop P uptake from a soil layer is sensitive to crop P demand and the amounts of labile P, soil water, and roots in the layer. Stover and root P are added to the fresh organic P pool upon their death and/or incorporation into the soil. The decomposition of fresh and stable organic matter may result in a net immobilization of labile P or net mineralization of organic P.

Soil organic matter and N transformations closely follow P flows and are based on Seligman and Van Keulen's (1981) PAPRAN model. They are divided into fresh pools, consisting of decomposing crop residue and microbial biomass N, and stable pools, consisting of organic matter humus and humus N. The nutrient models are sensitive to the soil chemical and physical properties, nutrient requirements of crops, tillage practice, fertilizer rate, soil temperature, and soil water content.

Regression equations were developed to estimate labile P from soil test P, organic P from total N or organic C, and fertilizer availability index from soil chemical and taxonomic characteristics (Sharpley et al., 1984). Thus, except for mineral N and labile P, the minimum data set required to run the N and P models can be obtained from soil survey information.

C. Nutrient Availability

A user-oriented model, Decide, predicting the effects of soil P buffering capacity on residual fertilizer P effectiveness, was developed by Bennett and Ozanne (1972) and subsequently modified by Helyar and Godden (1976). Bennett and Bowden (1976) described Decide as a response curve prediction method of making fertilizer recommendations, where the response curve chosen is the exponential or Mitscherlich curve:

$$Y = A(1 - B\exp[-CX]) \tag{3}$$

where Y is yield per unit area, A is the maximum yield per unit area, and B is the relative response to the applied P. The rate of P applied (X) is standardized to $\mathrm{kg\,P\,ha^{-1}}$ and the curvature coefficient or buffer capacity (C) has reciprocal dimensions to X. The coefficients are adjusted for soil type, past soil fertility, and yield goals. The Decide model (Bennett and Ozanne, 1972) is an excellent example of an attempt to produce objective advice on the use of fertilizer P. It attempts to reconstruct a fertilizer response curve for each farming situation, by combining the accumulated experience of research workers with the farmer's knowledge of his own farming conditions, in terms of past soil fertility and his future yield goals. The model was developed using highly weathered and leached soils of western Australia and is presently being modified for use on slightly weathered soils of eastern Australia. Consequently, the model only estimates fertilizer P recommendations for a limited range of soil types.

The data requirements, objectives, and limitations of the four types of models outlined above are summarized in Table 5. It is clear that model application depends on a clear definition of the specific objectives in evaluating the problem (absolute values or relative comparisons). Following this, the space (field, farm, county, or state) and time scale (daily

Table 5. Data requirements, objectives, and limitations of four model types.

Model	Data requirements: Soil	Data requirements: Plant	Objectives	Limitations
Nutrient Uptake: Mechanistic				
Claassen-Barber	Diffusion coefficient Solution concentration Buffer capacity	Michalis-Menton constant Water influx Root radius, length and growth	Mechanistic—accounts for rhizosphere effects	Transformation between soil nutrient forms controlling solution Concentration not considered
Nutrient Cycling: Research Oriented				
Century	Detailed		Holistic—long-term management effects on soil nutrient cycling uptake	Detailed soil inorganic and organic nutrient information needed
Nutrient Cycling: Simplified				
EPIC	Soil survey data Soil NO_3 Soil test P	Generalized root depth and growth parameters	Same as above	Supply and demand to nutrient uptake. Nutrients in solution phase not considered
Nutrient Availability: User Oriented				
Decide	Residual P Buffer capacity Cost of fertilizer	Desired crop yield Price of product	User orientated-fertilizer-use recommendations	Applicability to wide range of soil types

single event, monthly, or long-term annual averages) of the problem should be defined. Probably the most important factor, however, is the data needs of the model in relation to what data are available to the user. A model may be selected but the user may not have the resources to collect required parameterization data. It should be made clear that a model's output is only as reliable as the data input.

D. Improving Modeling Efforts

To increase efficiency and practicality of model application, more attention should be directed to the collection of input data which have the greatest effect on the root extraction of nutrients. Sensitivity analysis of the Claassen-Barber model by Silberbush and Barber (1983) indicated that root growth rate and root radius were the most sensitive parameters influencing P uptake. Further, predicted P uptake was more sensitive to soil solution concentration than diffusion coefficient and buffer power. A similar sensitivity analysis of the EPIC model by Favis-Mortlock and Smith (1990) showed that soil temperature, moisture, and nutrient fixation factors most greatly influence predictive accuracy. It is suggested that these plant and soil parameters will exert an even greater influence on the predicted root extraction of nutrients for soils having low available nutrient levels compared to nutrient sufficient or excess soils.

The successful application of any models simulating the long-term effects of management on the root extraction of nutrients will depend to a large extent on the availability of input data and, in particular, soil nutrient status and availability. Information on soil physical and chemical properties can be obtained from large, compiled databases and/or FAO-UNESCO Soil Maps of the World (FAO UNESCO, 1971, 1974, 1977) for wide geographic areas. These databases and soil maps, however, are essentially a record of soil properties with little quantitative information on the parameters directly relevant to nutrient uptake models. In this regard, soil classification has an important role to play, particularly if it can be extended to provide information on the fertility and production potential of soils.

The United States system of soil classification divides soils into groups based on naturally occurring assemblages of properties (Soil Survey Staff, 1975). In many cases, however, these groups do not refer to any particular use or management of the soil, even though the classification system includes a large amount of potentially useful information. However, the United States soil classification system does present a basis for the transfer of agricultural technology based on soil information (Buol and Denton, 1984).

One attempt which uses the United States system of soil classification in agricultural technology transfer is the Fertility Capability Classification (FCC) system (Buol et al., 1975; Sanchez et al., 1982). The FCC is a technical system for grouping soils according to the types of problems they present for agronomic management of their chemical and physical properties. It emphasizes the quantifiable topsoil parameters as well as subsoil properties directly relevant to plant growth. The FCC classes indicate the main fertility-related soil constraints, which can be interpreted in relation to specific farming systems or land utilization types (Sanchez et al., 1982).

At the present time, soil nutrient availability is inferred from soil chemical and physical properties and no direct measurement of the likely fate of fertilizer nutrients or their effect on nutrient uptake is included in the classification system. If estimates of nutrient status and the likely fate of fertilizer nutrients, here after referred to as "fertility indices," could be integrated into the United States soil classification system, then the potential exists for substantially extending the usefulness of the classification system.

Recent research by the Benchmark Soils Project investigated the N and P fertilizer response of soils in the same family, at widely scattered sites in several tropical countries, to determine whether the response was the same at all locations within a single soil family (Beinroth et al., 1980). It was concluded that responses between locations were similar, so long as climatic variables and initial soil nutrient levels were determined. Consequently, the use of the proposed fertility indices, in conjunction with soil classification, to transfer agricultural technology appears to be feasible.

A phosphorus retention test, developed originally for determining anion exchange capacity in soils, provides an indication of the likely fate of fertilizer P in soils. Such a method is finding increasing use as a tool in classifying allophanic soils (Blakemore et al., 1981; Blakemore, 1984) and those of lower P retention capacity (Bache and Williams, 1971). As routine procedures, P and K retention are useful fertility indices which could be used with little further development other than field evaluation.

There is an increased awareness of the importance of S in crop production, particularly with forage and grain legumes which have a high S requirement. Although acid soils, having significant positive charge through the presence of oxides and hydrous metal oxides, are expected to retain sulfate against leaching, intensive cropping rapidly depletes the native sulfate present. Relatedly, the addition of lime decreases the positive charge with a consequent decrease in sulfate sorption (Marsh et al., 1987). There is, therefore, a need for the development of fertility indices for S, which assess sulfate status and the likely retention of added sulfate against leaching. It is also possible that a sulfate retention test could be useful in estimating the positive charge in soils, or at least in separating soils with a dominant positive charge from those with a dominant negative charge.

In light of the preceding discussion, several research needs are indicated for future model development and application:

a. better description of nutrient release to solution and availability of soil inorganic and organic forms, using combined kinetic and equilibrium approaches;
b. better description of microbial processes and intermediary pools of plant nutrients;
c. better description of nutrient interaction effects;
d. better description of crop residue breakdown and organic matter incorporation and their effect on soil tilth;
e. improved linkage between soil and crop models;
f. better description of the interaction between crop roots and shoots, in terms of nutrient uptake, translocation, and growth;
g. development of databases of soil profile information (physical, chemical, and biological);

h. and most importantly, a better description of root development and soil exploration.

V. Conclusions

Although this discussion is not all-inclusive, it attempts to highlight the current state of knowledge and research needs concerning the impact of soil management on the root extraction of nutrients. It is clear that manageable variables controlling soil-water content, nutrient availability, root growth and development, and thereby, root extraction of nutrients are interactive, complex, and dynamic. There is, therefore, a need for team research involving physical, chemical, and biological disciplines, which should adopt a holistic approach to the management of soil-crop systems to maximize the root extraction of nutrients.

Major research priorities should focus on tillage practice, fertilizer type and placement, residue management, irrigation, and crop selection to coincide the positional availability of soil water and nutrients with periods of active root growth and nutrient uptake. For example, the integrated use of inorganic and organic amendments to enhance the efficient use of novel and existing fertilizer formulations should be evaluated for various tillage systems. A better understanding of water and nutrient movement, as well as root growth and nutrient uptake in soil, is needed to maximize the effectiveness of fertigation and specialized irrigation techniques for various cropping systems. In terms of crop selection, it is necessary to evaluate ways to increase the diversity of our cropping systems through alternative crops, alternative uses of forage crops, and conservation crops for diverted acres. Research should also explore ways to expand the use of cover crops through innovative establishment methods and selection of germplasm specifically "tailored" for cover crops.

It is also necessary to determine the relative importance of soil properties on organic matter cycling as influenced by soil fauna and flora, tillage, and crop rotation. Model development should be encouraged, keeping in mind the user's ultimate needs, data availability, and accuracy required. This development should include a better description of microbial processes influencing the rate of nutrient cycling in organic and inorganic pools and, most importantly, a better description of root development and soil exploration.

As mentioned at the beginning of this chapter, one of the main challenges facing agricultural research is to identify management practices that maintain long-term soil fertility and crop production with reduced chemical inputs and tillage operations. Maximizing the root extraction of indigenous and applied nutrients through appropriate soil and crop management outlined above will be of prime importance in meeting this challenge. Information from identified research needs and integration of existing knowledge into management systems will facilitate this progress.

Acknowledgment

The assistance of Elaine Mead in preparing the text and graphics for this manuscript is gratefully acknowledged.

References

Abbott, J.L. 1978. Importance of the organic phosphorus fraction in extracts of calcareous soils. *Soil Sci. Soc. Am. J.* 42:81–85.

Adams, F. 1981. Alleviating chemical toxicities: Liming acid soils. *In* G.F. Arkin and H.M. Taylor (Eds.). *Modifying the Root Environment to Reduce Crop Stress*. Am. Soc. Agric. Eng. Monogr. 4, St. Joseph, Michigan, pp. 269–301.

Adams, F. 1984. Crop response to lime in the southern U.S. *In* F. Adams (Ed.). *Soil Acidity and Liming, 2nd Ed.* Am. Soc. Agron. Monogr. No. 12, Madison, Wisconsin, pp. 211–266.

Adams, F., and Z.F. Lund. 1966. Effect of chemical activity of soil solution aluminum on cotton root penetration of acid subsoils. *Soil Sci.* 101:193–198.

Adams, F., and R.W. Pearson. 1970. Differential response of cotton and peanuts to subsoil acidity. *Agron. J.* 62:9–12.

Adepetu, J.A., and R.B. Corey. 1976. Organic phosphorus as a predictor of plant available phosphorus in soils of Southern Nigeria. *Soil Sci.* 122:159–164.

Ahuja, L.R. 1986. Characterization and modeling of chemical transfer to runoff. *Adv. Soil Sci.* 4:149–188.

Alison, M.W., and C.S. Hoveland. 1989. Root and herbage growth response of bird's-foot trefoil entries to subsoil acidity. *Agron. J.* 81:677–680.

Anghioni, I., and S.A. Barber. 1980. Predicting the most efficient phosphorus placement for corn. *Soil Sci. Soc. Am. J.* 44:1016–1020.

Apthorp, J.N., M.J. Hedley, and R.W. Tillman. 1987. The effect of nitrogen fertilizer form on the plant availability of phosphate from soil, phosphate rock, and monocalcium phosphate. *Fert. Res.* 12:269–284.

Arminger, W.H., C.D. Foy, A.L. Fleming, and B.E. Caldwell. 1968. Differential tolerance of soybean varieties to an acid soil high in exchangeable aluminum. *Agron. J.* 60:67–70.

Asea, P.E.A., R.M.N. Kucey, and J.W.B. Stewart. 1988. Inorganic phosphate solubilization by two *Penicillium* species in solution culture and soil. *Soil Biol. Biochem.* 20:459–464.

Bache, B.W., and E.G. Williams. 1971. A phosphate sorption index for soils. *Soil Sci.* 22:289–301.

Bailey, L.D., and C.A. Grant. 1989. Fertilizer phosphorus placement studies on calcareous and noncalcareous chernozemic soils: Growth, P-uptake and yield of flax. *Commun. Soil Sci. Plant Anal.* 20:635–654.

Banik, S., and B.K. Dey. 1982. Available phosphate content of an alluvial soil as influenced by inoculation of some isolated phospate-solubilizing microorganisms. *Plant Soil* 69:353–364.

Barber, S.A., A.D. Mackay, R.O. Kuchenbuch, and P.B. Barraclough. 1988. Effects of soil temperature and water on maize root growth. *Plant Soil* 111: 267–269.

Barea, J.M., and C. Azcon-Aguilar. 1983. Micorrhizas and their significance in nodulating nitrogen-fixing plants. *Adv. Agron.* 36:1–54.

Barley, K.P. 1959. The influence of earthworms on soil fertility. I. Earthworm populations found in agricultural land near Adelaide. *Aust. J. Agric. Res.* 10:171–178.

Barley, K.P., and A.C. Jennings. 1959. Earthworms and soil fertility. III. The influence of earthworms on the availability of nitrogen. *Aust. J. Agric. Res.* 10:364–370.

Barnes, J.S., and E.J. Kamprath. 1975. Availability of North Carolina rock phosphate applied to soils. *North Carolina Agric. Exp. Sta. Tech. Bull. 229*, 23 pp.

Barrow, N.J., and E.D. Carter. 1978. A modified model for evaluating residual phosphate in soil. *Aust. J. Agric. Res.* 29:1011–1021.

Beinroth, F.H., U. Uehara, J.A. Silva, R.W. Arnold, and F.B. Cady. 1980. Agrotechnology transfer in the tropics based on soil taxonomy. *Adv. Agron.* 33:303–339.

Bennett, D., and J.W. Bowden. 1976. "Decide"—An aid to efficient use of phosphorus. *In* G.J. Blair (Ed.). *Prospects for Improving Efficiency of Phosphorus Utilization.* Proc. Symp. University New England, Reviews in Rural Sci. III, Armidale, N.S.W., Australia, pp. 77–81.

Bennett, D., and P.G. Ozanne. 1972. Australia, CSIRO Division of Plant Industry. *Annual Report*, 45 pp.

Bernstein, L. 1964. Salt tolerance of plants. *United States Dept. Agric. Info. Bull. 283*. 23 pp.

Bernstein, L., L.E. Francois, and R.A. Clark. 1974. Interactive effects of salinity and fertility on grains and vegetables. *Agron. J.* 66:412–421.

Bertilsson, G. 1988. Lysimeter studies of nitrogen leaching and nitrogen balances as affected by agricultural practices. *Acta Agric. Scand.* 38:3–11.

Black, A.L., P.L. Brown, A.D. Halvorson, and F.M. Siddoway. 1981. Dry land cropping strategies for efficient water-use to control saline seeps in the northern Great Plains, U.S.A. *Agric. Water Manage.* 4:295–311.

Blackmer, A.M., D. Pottker, M.E. Cerrato, and J. Webb. 1989. Correlations between soil nitrate concentrations in late spring and corn yields in lowa. *J. Prod. Agric.* 2:103–109.

Blakemore, L.C. 1984. A field office method of phosphate retention. *New Zeal. J. Sci.* 27:33–36.

Blakemore, L.C., P.L. Searle, and B.K. Daly. 1981. Soil Bureau Laboratory methods. A method for chemical analysis of soils. *New Zeal. Soil Bureau Scientific Report*, 10A.

Blal, B., C. Morel, V. Gianinazzi-Pearson, J.C. Fardeau, and S. Gianinazzi. 1990. Influence of vesicular-arbuscular mycorhizae on phosphate fertilizer efficiency in two tropical acid soils planted with micropropagated soil palm (*Elaeis guineensis* jacq.) *Bio. Fert. Soils* 9:43–48.

Bolan, N.S., M.J. Hedley, and R.E. White. 1989. Nitrogen fertilizer use, fixation and soil acidification. *In* R.E. White and L.D. Currie (Eds.), *Nitrogen in New*

Zealand Agriculture and Horticulture. Fertilizer and Lime Research Centre, Massey University, Palmerston North, New Zealand, pp. 88–103.

Bolan, N.S., A.D. Robson, and N.J. Barrow. 1983. Plant and soil factors including mycorrhizal infection causing sigmoidal response of plants to applied phosphorus. *Plant Soil* 73:187–203.

Bolan, N.S., A.D. Robson, and N.J. Barrow. 1987. Effects of phosphorus application and mycorrhizal inoculation on root characteristics of subclover and ryegrass in relation to phosphorus uptake. *Plant Soil* 104:294–298.

Borkert, C.M., and S.A. Barber. 1985. Soybean shoot and root growth and phosphorus concentration as affected by phosphorus placement. *Soil Sci. Soc. Am. J.* 49:152–155.

Bouton, J.H., J.E. Hammel, and M.E. Sumner. 1982. Alfalfa, *Medicago sativa* L., in highly weathered acid soils. IV. Root growth into acid subsoil of plants selected for acid tolerance. *Plant Soil* 65:187–192.

Bowen, H.D. 1981. Alleviating mechanical impedance. *In* G.F. Arkin and H.M. Taylor (Eds.). Modifying the Root Environment to Reduce Crop Stress. Am. Soc. Agric. Eng. Monogr. 4, St. Joseph, Michigan, pp. 21–57.

Bowman, R.A., and C.V. Cole. 1978. Transformations of organic phosphorus substances in soils as evaluated by $NaHCO_3$ extraction. *Soil Sci.* 125:49–53.

Brenes, E., and R.W., Pearson. 1973. Root responses of three gramineal species to soil acidity in an oxisol and an ultisol. *Soil Sci.* 116:295–302.

Buol, S.W., and H.P. Denton. 1984. The role of soil classification in technology transfer. *In* R.B. Grossman, R.H. Rust, and H. Eswaran (Eds.). *Soil Taxonomy—Achievements and Challenges*. Am. Soc. Agron. Monogr., Madison, Wisconsin, pp. 29–43.

Buol, S.W., P.A. Sanchez, R.B. Cate, Jr., and M.A. Granger. 1975. Soil fertility capability classification. *In* E. Bornemisza and A. Alaraado (Eds.). *Soil Management in Tropical America*. Soil Sci. Dept., North Carolina State University, Raleigh, N.C., pp. 126–141.

Buwalda, J.G., D.P. Stribley, and P.B. Tinker. 1983. Increased uptake of anions by plants with vesicular-arbuscular mycorrhizae. *Plant Soil* 71:463–467.

Cabala-Rosand, P., and A. Wild. 1982. Direct use of low grade phosphate rock from Brazil as fertilizer. II. Effects of mycorrhizal inoculation and nitrogen source. *Plant Soil* 65:363–373.

Cabrera, M.L., and D.E., Kissel. 1988. Potentially mineralizable nitrogen in disturbed and undisturbed soil samples. *Soil Sci. Soc. Am. J.* 52:1010–1015.

Campbell, C.A., R. DeJong, and K.P. Zenther. 1984. Effect of cropping, summer fallow, and fertilizer N on nitrate-N lost by leaching on a Brown Chernozemic loam. *Can. J. Soil Sci.* 64:61–74.

Cannell, R.Q., and M.B. Jackson. 1981. Alleviating aeration stress. *In* G.F. Arkin and H.M. Taylor (Eds.). *Modifying the Root Environment to Reduce Crop Stress*. Am. Soc. Agric. Eng. Monogr. 4, St. Joseph, Michigan, pp. 141–192.

Carter, M.R., and D.A. Rennie. 1982. Changes in soil quality under zero tillage farming systems: Distribution of microbial biomass and mineralizable C and N potentials. *Can J. Soil Sci.* 62:586–597.

Cerda, A., F.T. Bingham, and G.J. Hoffman. 1977. Interactive effect of salinity and phosphorus on sesame. *Soil Sci. Soc. Am. J.* 41:915–918.

Champagnol, F. 1979. Relationship between phosphate nutrition of plants and salt toxicity. *Phosphorus in Agriculture* 76:35–44.

Chaudhary, T.N., V.K. Bhathagar, and S.S. Prihar. 1974. Growth response of crops to depth and salinity of ground water, and soil submergence. I. Wheat. *Agron. J.* 66:32–35.

Chaudhary, T.N., V.K. Bhathagar, and S.S. Prihar. 1975. Corn yield and nutrient uptake as affected by water-table depth and soil submergence. *Agron. J.* 67:745–749.

Chien, S.H., and D.K. Friesen. 1992. Phosphate rock for direct application. *In* F.J. Sikora (Ed.). *Future Directions for Agricultural Phosphorus Research.* National Fert. and Environ. Res. Cent., TVA, Muscle Shoals, Alabama, 18 pp.

Chien, S.H., F. Adams, F.E. Khasawneh, and J. Henoa. 1987a. Effects of combination of TSP and a reactive phosphate rock on yield and phosphorus uptake by corn. *Soil Sci. Soc. Am. J.* 51:1656–1658.

Chien, S.H., D. Sompongse, J. Henao, and D.T. Hellums. 1987b. Greenhouse evaluation of phosphorus availability from compacted phosphate rocks with urea or with urea and triple superphosphate. *Fert. Res.* 14:257–263.

Claassen, N., and S.A. Barber. 1974. A method for characterizing the relation between nutrient concentration and flux into roots of intact plants. *Plant Physiol.* 54:564–568.

Claassen, N., and S.A. Barber. 1976. Simulation model for nutrient uptake from soil by a growing plant root system. *Agron. J.* 68:961–964.

Cole, C.V., G.S. Innis, and J.W.B. Stewart. 1977. Simulation of phosphorus cycling in semi-arid grasslands. *Ecology* 58:1–15.

Cooper, R. 1959. Bacterial fertilizers in the Soviet Union. *Soils Fert.* 22:327–333.

Cooperband, L.R., and T.J. Logan. 1991. Measuring in situe changes in labile soil P in a humid tropical silvopastoral system with anion exchange membranes. *Agron. Abstracts*, 241.

Cox, F.R., E.J. Kamprath, and R.E. McCollum. 1981. A descriptive model of soil test nutrient levels following fertilization. *Soil Sci. Soc. Am. J.* 45:529–532.

Cruse, R.M., G.A. Yable, T.C. Colvin, D.R. Timmons, and A.C. Mussleman. 1983. Tillage effects on corn and soybean production in farmer-managed, university monitored field plots. *J. Soil Water Conserv.* 38:512–514.

Crush, J.R. 1973. The effect of *Rhizophagus tenius* mycorrhizas on rye grass, cocksfoot, and sweet vernal. *New Phytol.* 72:965–973.

Currie, J.A. 1970. Movement of gases in soil respiration. *In Sorption and Transport Processes in Soils*. Soc. Chem. Ind. Monogr. 37:152–171.

Cushman, J.H. 1982. Nutrient transport inside and outside the root rhizosphere: Theory. *Soil Sci. Soc. Am. J.* 46:704–709.

Cushman, J.H. 1984. Nutrient transport inside and outside the root rhizosphere: Generalized model. *Soil Sci.* 138:164–171.

Daughtrey, Z.W., J.W. Gilliam, and E.J. Kamprath. 1973. Soil test parameters for assessing plant-available P of acid organic soils. *Soil Sci.* 115:438–446.

De Boer, D.W., and W.F. Ritter. 1970. Flood damage to crops in depressional areas of northcentral lowa. *Trans. Am. Soc. Agric. Eng.* 13:549 and 553.

Dixon, R.M., and A.E. Peterson. 1971. Water infiltration control. A channel system concept. *Soil Sci. Soc. Am. Proc.* 35:968–973.

Doran, J.W. 1980. Soil microbial and biochemical changes associated with reduced tillage. *Soil Sci. Soc. Am. J.* 44:765–771.

Doran, J.W., and M.S. Smith. 1987. Organic matter management and utilization of soil and fertilizer nutrients. *In* R.F. Follett, J.W.B. Stewart, and C.V. Cole (Eds.). *Soil Fertility and Organic Matter as Critical Components of Production Systems*. Spec. Publ. No. 19, Soil Sci. Soc. Am., Madison, Wisconsin, pp. 53–72.

Dregne, H.E., and H. Mojallali. 1969. Salt-fertilizer specific ion interactions in soil. New Mexico Agric. Ext. Sta. Bull. No. 541. 16 pp.

Dwyer, L.M., D.W. Stewart, and D. Balchin. 1988. Rooting characteristics of corn, soybeans, and barley as a function of available water and soil physical characteristics. *Can. J. Soil. Sci.* 68:121–132.

Eckert, D.J., and J.W. Johnson. 1985. Phosphorus fertilization in no-till corn production. *Agron. J.* 77:789–792.

Edwards, C.A., and J.R. Lofty. 1978. The influence of arthropods and earthworms upon root growth of direct drilled cereals. *J. Appl. Ecol.* 15:789–795.

Edwards. C.A., and J.R. Lofty. 1980. Effects of earthworm inoculations on the root growth of direct drilled cereals. *J. Appl. Ecol.* 17:533–544.

Edwards, W.M., M.J. Shipitalo, L.B. Owens, and L.D. Norton. 1989. Water and nitrate movement in earthworm burrows within long-term no-till cornfields. *J. Soil Water Conserv.* 44:240–243.

Elkins, C.B., P.L. Haaland, and C.S. Hoveland. 1977. Grass roots as a tool for penetrating soil hardpans and increasing crop yields. *Proc. 34th Southern Pasture Forage Crop Imp. Conf.* Auburn, Alabama, pp. 21–26.

Ellis, J.R. 1988. Effects of tillage on VAM colonization of wheat in Nebraska. *In* J. Thompson (Ed.). *Vesicular-Arbuscular Mycorrhizae in Agriculture and Horticulture*. Proc. Aust. Instit. Agric. Res., Toowoomba, Queensland, Australia, pp. 47–48.

Ellis, J.R. 1990. The role of mycorrhizal colonization of corn in decreasing nitrate leaching in soil columns. *Agron. Abstr.* 1990:248.

Ellis, J.R., M.E. Busse, and S.D. Masterson. 1988. Effects of VAM on atrazine uptake by soybean. *In* J. Thompson (Ed.). *Vesicular-Arbuscular Mycorrhizae in Agriculture and Horticulture*. Proc. Aust. Instit. Agric. Res., Toowoomba, Queensland, Australia, pp. 14–15.

Ellis, J.R., H. Larsen, and M. Boosalis. 1985. Drought resistance of wheat plants inoculated with vesicular-arbuscular mycorrhizae. *Plant Soil* 86:369–378.

Ellis, F.B., J.G. Elliot, B.T. Barnes, and K.R. Howse. 1977. Comparison of direct drilling, reduced cultivation and ploughing on the growth of cereals. 2. Spring barley on a sandy loam soil: Soil physical conditions and root growth. *J. Agric. Sci.* 89:631–642.

Evans, D.G., and M.H. Miller. 1988. Vesicular-arbuscular mycorrhizas and the soil-disturbance-induced reduction of nutrient absorption in maize. I. Causal relations. *New Phytologist* 110:67–74.

Fairchild, M.H., and G.L. Miller. 1988. Vesicular-arbuscular mycorrhizas and the soil-disturbance-induced reduction of nutrient absorption in maize. II. Development of the effect. *New Phytologist* 110:75–84.

FAO UNESCO. 1971. *Soil Map of the World. 4. South America*. UNESCO, Paris, France.

FAO UNESCO. 1974. *Soil Map of the World. 7. Legend*. UNESCO, Paris, France.

FAO UNESCO. 1977. *Soil Map of the World. 6. Agrica*. UNESCO, Paris, France.

Farina, M.P.W., and P. Channon. 1988. Acid-subsoil amelioration. II. Gypsum effects on growth and subsoil chemical properties. *Soil Sci. Soc. Am. J.* 52: 175–180.

Fausey, N.R., T. Van Toai, and M.B. McDonald. 1983. Response of ten corn cultivars to flooding. Am. Soc. Agric. Eng. Winter Mtg., Chicago, Illinois Dec. 13–16 Paper No. 83-2558. Am. Soc. Agric. Eng., St. Joseph, Michigan.

Favis-Mortlock, D.T., and F.R. Smith. 1990. A sensitivity analysis of EPIC. *In* A.N. Sharpley and J.R. Williams (Eds.). *EPIC—Erosion-Productivity Impact Calculator. I. Model Documentation*. United States Dept. Agric. Tech. Bull. 1768, Govt. Printing Office, Washington, D.C., Chapter 9, pp. 178–190.

Feigin, A. 1985. Fertilization management of crops irrigated with saline water. *Plant Soil* 89:285–299.

Fine, L.O., and P.L. Carson. 1954. Phosphorus needs of small grains on a moderately saline soil. *Soil Sci. Soc. Am. Proc.* 18:60–63.

Fink, R.J., and D. Wesley. 1974. Corn yield as affected by fertilization and tillage system. *Agron. J.* 66:637–639.

Follett, R.F., and G.A. Peterson. 1988. Surface soil nutrient distribution as affected by wheat-fallow tillage systems. *Soil Sci. Soc. Am. J.* 52:141–147.

Foy, C.D. 1974. Effects of aluminum on plant growth. *In* E.W. Carson (Ed.). *The Plant Root and Its Environment*. University Press, Charlottesville, Virginia, pp. 601–642.

Foy, C.D. 1984. Physiological effects of hydrogen, aluminum, and manganese toxicities in acid soil. *In* F. Adams (Ed.). *Soil Acidity and Liming, 2nd Ed.* Am. Soc. Agron. Monogr. No. 12, Madison, Wisconsin, pp. 57–97.

Francois, L.E., T.J. Donovan, and E.V. Maas. 1984. Salinity effects on seed yield, growth, and germination of grain sorghum. *Agron J.* 76:741–744.

Francois, L.E., T.J. Donovan, K. Lorenz, and E.V. Maas. 1980. Salinity effects on rye grain yield, quality, vegetative growth, and emergence. *Agron. J.* 81:707–712.

Francois, L.E., T.J. Donovan, E.V. Maas, and G.L. Rubenthaler. 1988. Effect of salinity on grain yield and quality, vegetative growth and germination of triticale. *Agron. J.* 80:642–647.

Francois, L.E., E.V. Maas, T.J. Donovan, and V.L. Youngs. 1986. Effect of salinity on grain yield and quality, vegetative growth, and germination of semi-dwarf and durum wheat. *Agron. J.* 78:1053–1058.

Gerdemann, J.W. 1968. Vesicular-arbuscular mycorrhiza and plant growth. *Ann. Rev. Phytopathol.* 6:397–418.

Germann, P.E.W., W.M. Edwards, and L.B. Owens. 1984. Profiles of bromide and increased soil moisture after infiltration into soils with macropores. *Soil Sci. Soc. Am. J.* 48:237–244.

Gianello, C., and J.M. Bremner. 1988. A rapid steam distillation method of assessing potentially available organic nitrogen in soil. *Commun. Soil Sci. Plant Anal.* 19:1551–1568.

Gianinazzi-Pearson, V. 1986. Mycorrhiza: A potential for better use of phosphate fertilizer. *Fertilizers and Agriculture* 92:3–12.

Gilbertson, C.B., F.A. Norstadt, A.C. Mathers, R.F. Holt, A.P. Barnett, T.M. McCalla, C.A. Onstad, and R.A. Young. 1979. Animal waste utilization on cropland and pastureland. *United States Dept. Agric. Utilization Research Report No. 6*. EPA-600/2-79-059. 135 pp.

Gill, W.R., and G.H. Bolt. 1955. Pfeffer's studies of the root growth pressure exerted by plants. *Agron. J.* 47:166–168.

Gillespie, A.R., and P.E. Pope. 1990. A comparison of three rock phosphates for use in forest soils. *Fert. Res.* 23:141–146.

Gorbenko, A.Y., N.S. Panikov, and D.V. Zvyaginstev. 1986. The effect of invertebrates on the growth of soil microorganisms. *Mikrobiologia* 55: 515–521.

Grable, A.R. 1966. Soil aeration and plant growth. *Adv. Agron.* 18:57–106.

Graff, O. 1970. The phosphorus content of earthworm casts. *LandbForsch-Volkenrode* 20:33–36.

Grattan, S.R., and E.V. Maas. 1984. Interactive effects of salinity and substrate phosphate on soybean. *Agron. J.* 76:668–676.

Grattan, S.R., and E.V. Maas. 1985. Root control of leaf phosphorus and chlorine accumulation in soybean under salinity stress. *Agron. J.* 77:890–895.

Grecu, S.J., M.B. Kirkham, E.T. Kanemasu, D.W. Sweeney, L.R. Stone, and G.A. Milliken. 1988. Root growth in a claypan with a perennial-annual rotation. *Soil Sci. Soc. Am. J.* 52:488–494.

Gregory, P.J. 1988. Growth and functioning of plant roots. *In* A. Wild (Ed.). *Russell's Soil Conditions and Plant Growth*. John Wiley and Sons, Inc., New York, pp. 113–167.

Guild, W.J. 1955. Earthworms and soil structure. *In* D.K. McE. Kevan (Ed.). *Soil Zoology*. Butterworth, London, England, pp. 83–98.

Gupta, V.V.S.R., and J.J. Germida. 1988. Distribution of microbial biomass and its activity in different soil aggregate size classes as affected by cultivation. *Soil Biol. Biochem.* 20:777–786.

Guttay, A.J.R., and L.M.C. Dandurand. 1989. Interaction of the vesicular-arbuscular mycorrhizae of maize with extractable soil phosphorus levels and nitrogen-potassium fertilizers. *Biol. Fert. Soils* 8:307–310.

Halvorson, A.D., B.R. Bock, and J.J. Meisinger. 1989. Advances in N use efficiency. *In* G.H. Heichel, R.F. Follett, J.F. Power, and D.R. Keister (Eds.). Current *Advances and Future Priorities in Nitrogen Research*. USDA-ARS Tech. Rept., Govt. Printing Office, Washington, D.C. 16 pp.

Hamm, J.W., J.R. Bettany, and E.H. Halstead. 1973. A soil test for sulfur and interpretive criteria for Saskatchewan. *Commun. Soil Sci. Plant Anal.* 4:219–231.

Hammel, J.E., M.E. Sumner, and H. Shahandel. 1985. Effect of physical and chemical profile modification on soybean and corn production. *Soil Sci. Soc. Am. J.* 49:1508–1511.

Hammond, L.L., S.H. Chien, and A.V. Mokwunye. 1986. Agronomic value of unacidulated and partially acidulated phosphate rocks indigenous to the tropics. *Adv. Agron.* 40:89–140.

Hansson, A.C., and O. Andren. 1987. Root dynamics in barley, lucerne, and meadow fescue investigated with a mini-rhizotron technique. *Plant Soil* 103:33–38.

Harley, J.L. 1989. The significance of mycorrhizae. *Mycol. Res.* 92:129–139.

Harrison, R.A. 1987. *Soil Organic Phosphorus. A Review of World Literature.* C.A.B. International, Wallingford, Oxon, United Kingdom. 257 pp.

Havelka, N.D., M.G. Boyle, and R.W.F. Hardy. 1980. Biological nitrogen fixation. *In* F.J. Stevenson (Ed.). *Nitrogen in Agricultural Soils.* Agronomy 22, Am. Soc. Agron., Madison, Wisconsin, pp. 365–422.

Havlin, J.L. 1992. Sampling and evaluating residual value of band applied phosphorus. *In* F.J. Sikora (Ed.). *Future Directions for Agricultural Phosphorus Research.* National Fed. Environ. Res. Cent., TVA, Muscle Shoals, Alabama, 9 pp.

Haynes, R.J. 1980. Competitive aspects of the grass-legume association. *Adv. Agron.* 33:227–261.

Hecht-Buchholz, C., and C.D. Foy. 1981. Effect of aluminum toxicity on root morphology of barley. *In* R. Brouwer, O. Casparikova, J. Kolek, and B.C. Loughman (Eds.). *Structure and Function of Plant Roots.* Martinus Nijhoff/Dr. W. Junk Publishers, The Hague, The Netherlands, pp. 343–345.

Hedley, M.J., R.W. Tillman, and G. Wallace. 1989. The use of nitrogen fertilizers for increasing the suitability of reactive phosphate rocks for use in intensive agriculture. *In* R.E. White and L.D. Currie (Eds.). *Nitrogen Fertilizer Use in New Zealand Agriculture and Horticulture.* Occasional Report No. 3, Fertilizer and Lime Research Centre, Massey University, Palmerston North, New Zealand.

Hedley, M.J., A. Hussin, and N.S. Bolan. 1990. New approaches to phosphate fertilization. pp. 125–142. *In* H. Tiessen and S.S. Banta (Eds.). *Phosphorus Requirements for Sustainable Agriculture in Asia and Oceania.* Scientific Committee on Problems of the Environment/United Nation Development Program. Published by Int. Rice Res. Inst., Manila, Phillipines.

Hedley, M.J., R.E. White, and P.H. Nye. 1982. Plant-induced changes in the rhizosphere of rape (*Brassica napus* var. Emerald) seedlings. II. Origin of the pH changes. *New Phytol.* 91:31–44.

Helyar, K.R., and D.P. Godden. 1976. The phosphorus cycle—What are the sensitive areas? *In* G.J. Blair (Ed.). *Prospects for Improving Efficiency of Phosphorus Utilization.* Proc. Symp. University New England, Armidale, N.S.W., Australia, Reviews in Rural Sci. III, pp. 23–30.

Helyar, K.R., and D.N. Munns. 1975. Phosphate fluxes in the soil-plant system: A computer simulation. *Hilgardia* 43:103–130.

Holford, I.C.R. 1989. Efficacy of different phosphate application methods in relation to phosphate sorptivity in soils. *Aust. J. Soil Res.* 27:123–133.

Hoogerkamp, M., H. Rogaar, and H.J.P. Eijsackers. 1983. Effect of earthworms on grassland on recently reclaimed polder soils in the Netherlands. *In* J.E. Satchel (Ed.). *Earthworm Ecology: From Darwin to Vermiculture.* Chapman and Hall, London, England, pp. 85–105.

Hook, J.E., and G.J. Gascho. 1988. Multiple cropping for efficient use of water and nitrogen. *In* W.L. Hargrove (Ed.). *Cropping Strategies for Efficient Use of Water and nitrogen.* Am. Soc. Agron. Spec. Pub. No. 51, ASA, CSSA, SSSA, Madison, Wisconsin, pp. 7–20.

Howeler, R.H., C.J. Asher, and D.G. Edwards. 1982. Establishment of an effective mycorrhizal association on cassava in flowing solution culture and its effects on phosphorus nutrition. *New Phytol.* 90:229–238.

Hoyt, P.B., and R.H. Leitch. 1983. Effect of forage legume species on soil moisture, nitrogen, and yield of succeeding barley crops. *Can. J. Soil Sci.* 62:125–136.

Huck, M.G. 1970. Variation in tap root elongation rate as influenced by composition of the soil air. *Agron. J.* 62:815–818.

Hue, N.V., F. Adams, and C.E. Evans. 1985. Sulfate retention by an acid BE horizon of an ultisol. *Soil Sci. Soc. Am. J.* 49:1196–1200.

Jayachandran, J., A.P. Schwab, and B.A.D. Hetrick. 1989. Mycorrhizal mediation of phosphorus availability. Synthetic iron chelate effects on phosphorus solubilization. *Soil Sci. Soc. Am. J.* 53:1701–1706.

Jones, C.A. 1983. Effect of soil texture on critical bulk densities for root growth. *Soil Sci. Soc. Am. J.* 47:1208–1211.

Jones, C.A., C.V. Cole, A.N. Sharpley, and J.R. Williams. 1984. A simplified soil and plant phosphorus model. I. Documentation. *Soil Sci. Soc. Am. J.* 48:800–805.

Kafkafi, U., J. Letey, and N. Valoras. 1982. Chloride interaction with nitrate and phosphate nutrition in tomato. *J. Plant Nutr.* 5:1369–1385.

Kamprath, E.J. 1984. Crop response to lime on soils in the tropics. *In* F. Adams (Ed.). *Soil Acidity and Liming, 2nd Ed.* Am. Soc. Agron. Monogr. No. 12, Madison, Wisconsin, pp. 349–368.

Karlen, D.L., and C.R. Camp. 1985. Row spacing, plant population, and water management effects on corn in the Atlantic coastal plain. *Agron. J.* 77:393–398.

Khalil, M.A., F. Amer, and M. Elgabaly. 1967. A salinity-fertility interaction study on corn and cotton. *Soil Sci. Soc. Am. Proc.* 31:683–686.

Khasawneh, F.E., and E.C. Doll. 1978. The use of phosphate rock for direct application to soils. *Adv. Agron.* 30:159–206.

Kohl, R.A., and J.J. Kolar. 1976. Soil water uptake by alfalfa. *Agron. J.* 68:536.

Kovar, J.L., and S.A. Barber. 1987. Placing phosphorus and potassium for greatest recovery. *J. Fert. Issues* 4:1–6.

Kucey, R.M.N. 1987. Increased phosphorus uptake by wheat and field beans inoculated with a phosphorus-solubilizing *Penicillium bilaii* strain and with vesicular arbusular mycorrhizal fungi. *Appl. Environ. Microbiol.* 53:2699–2703.

Kucey, R.M.N. 1992. Microbial phosphate solubilization: Summary of research with *Penicillium bilaii. In* F.J. Silora (Ed.). *Future Directions for Agricultural Phosphorus Research.* National Fert. Environ. Res. Cent., TVA, Muscle Shoals, Alabama, 9 pp.

Kucey, R.M.N., and M.E. Leggett. 1989. Increased yields and phosphorus uptake by westar canola (*Brassica rapus* L.) inoculated with a phosphate-solubilizing isolate of *Penicillium bilaji. Can. J. Soil Sci.* 69:425–432.

Kucey, R.M.N., H.H. Janzen, and M.E. Leggett. 1989. Microbially mediated increases in plant-available phosphorus. *Adv. Agron.* 42:199–228.

Kuchenbuch, R.O., and S.A. Barber. 1987. Yearly variation in root distribution with depth in relation to nutrient uptake and corn yield. *Commun. Soil Sci. Plant Anal.* 18:255–263.

Kuo, S. 1990. Phosphate sorption implications on phosphate soil tests and uptake by corn. *Soil Sci. Soc. Am. J.* 54:131–135.

Kurtz, L.T., L.V. Boune, T.R. Peck, and R.G. Hoeft. 1984. Crop rotations for efficient nitrogen use. *In* R.D. Hauck (Ed.). *Nitrogen in Crop Production.* ASA, CSSA, SSSA, Madison, Wisconsin, pp. 295–306.

Lammond, R.E. 1987. Comparison of fertilizer solution placement methods for grain sorghum under two tillage systems. *J. Fert. Issues* 4:43–47.

Lapeyrie, F. 1988. Oxalate synthesis from soil bicarbonate by the mycorrhizal fungus *Paxillus involutus*. *Plant Soil* 110:3–8.

Lavelle, P. 1978. Les vers de terre de la savane de Lamto (Cote d'lvoire). Peuplements, populations, et fonctions de Pecosysteme. *Publ. Lab. Zool.* 12:1–301.

Lavelle, P. 1988. Earthworm activities and the soil system. *Biol. Fert. Soils* 6:237–251.

Lee, K.E. 1983. The influence of earthworms and termites on soil nitrogen cycling. *In* P. Lebrun, H.M. Andre, A. De Medts, C. Gregoire-Wibo, and G. Wauthy (Eds.). *New Trends in Soil Biology*. Proc. 8th Intl. Coll. Soil Zool., Louvain-la-Neuve, 1982, Dieu-Brichart, Ottignies-Louvain-la-Neuve, Belgium, pp. 35–48.

Lee, K.E. 1985. *Earthworms: Their Ecology and Relationships with Soils and Land Use*. Academic Press, London, England.

Lewis, T. 1981. Earthworms and direct drilling. *Rothamsted Experimental Station Report for 1980*, Pt. 1, 101–102.

L'honeux, B. de., L. Gerlach, A. Clemente, M.L. Roda-Santos, and J.A.G. Menaia. 1988. Ultrastructural characterization of rice husk submitted to different pretreatments to optimize its fermentation. *Biological Wastes* 23: 163–180.

Lowry, F.E., H.M. Taylor, and M.G. Huck. 1970. Growth rate and yield of cotton as influenced by depth and bulk density of soil pans. *Soil Sci. Soc. Am. Proc.* 34:306–309.

Lunt, H.A., and G.M. Jacobson. 1944. The chemical composition of earthworm casts. *Soil Sci.* 58:367–376.

Lutz, J.A., G.W. Hawkins, and C.F. Genter. 1971. Differential response of corn inbreds and single crosses to certain properties of an acid soil. *Agron. J.* 63:803–805.

Lynch, J.M., and D.A. Wood. 1985. Controlled microbial degradation of lignocellulose: Basis for existing and novel approaches to composting. *In* J.K.R. Gasser (Ed.). *Composting of Agricultural and Other Wastes*. Elsevier Applied Science Publishers, London, England, pp. 183–194.

Maas, E.V., and G.J. Hoffman. 1977. Crop salt tolerance-Current assessment. *J. Irr. Drain. Div.*, ASCE 103 (IR2):115–134.

Mackay, A.D., and S.A. Barber. 1984. Soil temperature effects on root growth and phosphorus uptake by corn. *Soil Sci. Soc. Am. J.* 48:818–823.

Mackay, A.D., J.K. Syers, J.A. Springett, and P.E.H. Gregg. 1982. Plant availability of phosphate in superphosphate and a phosphate rock as influenced by earthworms. *Soil Biol. Biochem.* 14:281–287.

Magdoff, F.R., D. Ross, and J. Amadon. 1984. A soil test for nitrogen availability to corn. *Soil Sci. Soc. Am. J.* 48:1301–1304.

Mansell, G.P., J.K. Syers, and P.E.H. Gregg. 1981. Plant availability of phosphorus in dead herbage ingested by surface casting earthworms. *Soil Biol. Biochem.* 13:163–167.

Mansell, R.S., H.M. Selim, and J.G.A. Fiskell. 1977. Simulated transformations and transport of phosphorus in soil. *Soil Sci.* 124:102–109.

Marsh, K.B., R.W. Tillman, and J.K. Syers. 1987. Charge relationships of sulfate sorption by soils. *Soil Sci. Soc. Am. J.* 51:318–323.

Martin, A. 1980. Sulphus in air and deposited from air and rain over Great Britain and Ireland. *Environ. Pollut.* 1:177–193.

Marwaha, B.C. 1989. Rock phosphate holds the key to productivity in acid soils. *Fert. News* 34:23–29.

Mathers, A.C., J.D. Thomas, B.A. Stewart, and J.E. Herring. 1980. Manure and inorganic fertilizer effects on sorghum and sunflower growth on iron-deficient soil. *Agron. J.* 72:1025–1029.

Maynard, D.G., J.W.B. Stewart, and J.R. Bettany. 1982. Diagnosis of sulfur deficiency in plants and soils. *In Proc. Alberta Soil Science Workshop.* Edmonton, Alberta, Canada, pp. 207–228.

McClenahan, E.J., and R. Killorn. 1988. Relationship between basal corn stem nitrate N content at V6 growth stage and grain yield. *J. Prod. Agric.* 1:322–326.

McLean, E.O., and J.R. Brown. 1984. Crop response to lime in the midwestern United States. *In* F. Adams (Ed.). *Soil Acidity and Liming, 2nd Ed.* Am. Soc. Agron. Monogr. No. 12, Madison, Wisconsin, pp. 267–304.

Meisinger, J.J., P.R. Shipley, and A.M. Decker. 1990. Using winter cover crops to recycle nitrogen and reduce leaching. *Proc. Southern Regional Tillage Conf.* July 14–16, 1990, North Carolina State Spec. Bull. 90-1, Raleigh, North Carolina.

Menon, R.G., and S.H. Chien. 1990. Phosphorus availability to maize from partially acidulated phosphate rocks and phosphate rocks compacted with triple superphosphate. *Plant Soil* 127:123–128.

Menon, R.G., L.L. Hammond, and H.A. Sissingh. 1989. Determination of plant available phosphorus by the iron hydroxide-impregnated filter paper (P_i) soil test. *Soil Sci. Soc. Am. J.* 52:110–115.

Menon, R.G., S.H. Chien, L.L. Hammond, and B.R. Arora. 1990. Sorption of phosphorus by the iron oxide-impregnated filter paper (P_i soil test) embedded in soils. *Plant Soil* 126:287–294.

Mishra, B., P.K. Khanna, and B. Ulrich. 1979. A simulation model for organic phosphorus transformation in a forest soil ecosystem. *Ecol. Model.* 6:31–46.

Misra, B., P.K. Omanwar, N.P. Misra, and R.D. Sharma. 1986. Efficacy of various mixtures of rock phosphate with pyrite or superphosphate as P sources. *In* G.V. Kothandaraman, T.S. Manickam, and K. Natarajan (Eds.). *Proc. of Workshop on Rockphosphate in Agriculture.* Tamil Nadu Agricultural University, Coimbatore, India.

Moody, P.W., R.L. Aitken, B.L. Compton, and S. Hunt. 1988. Soil phosphorus parameters affecting phosphorus availability to, and fertilizer requirements of, Maize (*Zea mays*). *Aust. J. Soil Res.* 26:611–622.

Morra, M.J., and W.A. Dick. 1988. Atmospheric sulfur deposition to agricultural land in northeastern Ohio. *J. Environ. Qual.* 17:323–329.

Moschler, W.W., G.M. Shear, D.C. Martens, G.D. Jones, and R.R. Wilmouth. 1972. Comparative yield and efficiency of no-till and conventionally tilled corn. *Agron. J.* 64:229–231.

Mosse, B. 1973. Advances in the study of vesicular-arbuscular mycorrhiza. *Ann. Rev. Phytophathol.* 11:171–196.

Mosse, B., D.S. Hayman, and D.J. Arnold. 1973. Plant growth responses to vesicular-arbuscular mycorrhiza. V. Phosphate uptake by three plant species from P deficient soils labeled with ^{32}P. *New Phytol.* 72:809–815.

Muchovej, R.M.C., J.J. Muchovej, and V.H. Alvarez. 1989. Temporal relations of phosphorus fractions in an oxisol amended with rock phosphate and *Thiobacillus thiooxidans*. *Soil Sci. Soc. Am. J.* 53:1096–1100.

Mugnier, J., G. Jung, and J.L. Prioul. 1986. Methods of producing endomycorrhizian fungi with arbuscules and vesicles in vitro. *United States Patent No. 4,599,312*.

Muir, J., J.S. Boyce, F.C. Seim, P.N. Moscher, E.J. Diebert, and R.A. Olson. 1976. Influence of crop management practices on nutrient movement below the root zone in Nebraska soils. *J. Environ. Qual.* 5:255–259.

Mulligan, M.F., A.J.M. Smucker, and G.F. Safir. 1985. Tillage modifications of dry edible bean root colonization by VAM fungi. *Agron. J.* 77:140–144.

Nardi, S., G. Arnoldi, and G. Dell'Agnola. 1988. Release of the hormone-like activities from *Allolobophora rosen* (Sav.) and *Allolobophora caliginosa*(Sav.) feces. *Can. J. Soil Sci.* 65:563–567.

Neenan, M. 1960. The effects of soil acidity on growth of cereals with particular reference to the differential reaction of varieties thereto. *Plant Soil* 12:324–328.

Nielson, R.L. 1965. Presence of plant growth substances in earthworms demonstrated by paper chromatography and the Went pea test. *Nature* (London) 208:113–114.

Nieman, R.H., and R.A. Clark. 1976. Interactive effects of salinity and phosphorus nutrition on the concentrations of phosphate and phosphate esters in mature photosynthesizing corn leaves. *Plant Physiol.* 57:157–161.

Nobel, A.D., M.E. Sumner, and A.K. Alva. 1988. The pH dependency of aluminum phytotoxicity alleviation by calcium sulfate. *Soil Sci. Soc. Am. J.* 52:1398–1402.

Nye, P.H. 1979. Diffusion of ions and uncharged solutes in soils and soil clays. *Adv. Agron.* 31:225–272.

Nye, P.H., and F.H.C. Marriott. 1969. A theoretical study of the distribution of substances around roots resulting from simultaneous diffusion and mass-flow. *Plant Soil* 30:459–472.

O'Connor, G.A., W.A. Lindsay, and S.R. Olsen. 1971. Diffusion of iron and iron chelates in soil. *Soil Sci. Soc. Am. Proc.* 35:407–410.

Ogo, T., and S. Morikawi. 1965. Relationships between certain nitrogen fractions in leaf blade of crops and salt tolerance. *Shimone Agr. Coll. Bull. 13A*. pp. 5–9.

Olsen, S.R., and W.D. Kemper. 1967. Movement of nutrients to plant roots. *Adv. Agron.* 20:91–151.

Olsen, S.R., C.V. Cole, F.S. Watanabe, and L.A. Dean. 1954. Estimation of available phosphorus in soils by extraction with sodium bicarbonate. *United States Dept. Agric. Circ. 939*. U.S. Gov. Print. Off. Washington, D.C.

O'Toole, J.C., and W.C. Bland. 1987. Genotypic variation in crop plant root systems. *Adv. Agron.* 41:91–145.

Oyandel, C., and J. Rodriguez. 1977. Estimacion de la mineralizacion del n en el suelo. *Ciencia Investigacion Agraria* 4:33–44.

Papadopoulos, I., and V.V. Rendig. 1983. Interactive effects of salinity and nitrogen on growth and yield of tomato plants. *Plant Soil.* 73:47–57.

Parle, J.N. 1963. Microorganisms in the intestines of earthworms. *J. Gen. Microbiol.* 31:1–11.

Parton, W.J., J.W.B. Stewart, and C.V. Cole. 1988. Dynamics of C, N, P, and S in grassland soils. A model. *Biogeochemistry* 5:109–131.

Parton, W.J., D.W. Anderson, C.V. Cole, and J.W.B. Stewart. 1983. Simulation of soil organic matter formations and mineralization in semi-arid agroecosystems. *In* R.R. Lowrance, R.L. Todd, L.E. Asmussen, and R.A. Leonard (Eds.). Nutrient cycling in agricultural ecosystems. *College of Agric. Exp. Sta. Special Publ. 23.* University Georgia, Athens, pp. 533–550.

Parton, W.J., D.S. Schimel, C.V. Cole, and D.S. Ojima. 1987. Analysis of factors controlling soil organic matter levels in Great Plains grasslands. *Soil Sci. Soc. Am. J.* 51:1173–1179.

Payne, D., and P.J. Gregory. 1988. The soil atmosphere. *In* A. Wild (Ed.). *Russell's Soil Conditions and Plant Growth.* John Wiley and Sons, Inc., New York, pp. 298–315.

Peterson, C.M., C.B. Elkins, and P.A. Roberts. 1976. Root development of bahiagrass (*Paspalum notatum var. saurae*). Light microscope and SEM studies. *Abstr. Bot. Soc. Am.* AIBS Meeting, May 1976, New Orleans, Louisiana.

Peterson, G.A., and W.W. Frye. 1989. Fertilizer nitrogen management. *In* R.F. Follett (Ed.). *Nitrogen Management and Ground Water Protection. Developments in Agricultural and Managed-Forest Ecology, 21.* Elsevier Press, Amsterdam, The Netherlands, pp. 183–219.

Phillips, R.E., and D. Kirkham. 1962. Soil Compaction in the field and corn growth. *Agron. J.* 54:29–34.

Powell, C.L. 1979. Effect of mycorrhizal fungi on recovery of phosphate fertilizer from soil by ryegrass plants. *New Phytol.* 83:681–694.

Power, J.F., and F.E. Broadbent. 1989. Proper accounting for N in cropping systems. *In* R.F. Follett (Ed.). *Nitrogen Management and Ground Water Quality Protection. Development in Agricultural and Managed-Forest Ecology, 21.* Elsevier Press, Amsterdam, Holland, pp. 160–181.

Power, J.F., G.A. Reichman, and D.L. Grunes. 1961. The influence of phosphorus fertilization and moisture on growth and nutrient adsorption by spring wheat. II. Soil and fertilizer P uptake in plants. *Soil Sci. Soc. Am. Proc.* 25:210–213.

Power, J.F., W.W. Whilhelm, and J.W. Doran. 1986. Crop residue effects on soil environment and dryland maize and soybean production. *Soil Tillage Res.* 8:101–111.

Radke, J.K., R.W. Andrews, R.R. Janke, and S.E. Peters. 1988. Low-input cropping systems and efficiency of water and nitrogen use. *In* W.C. Hargrove (Ed.). *Cropping Strategies for Efficient Use of Water and Nitrogen.* Am. Soc. Agron. Spec. Pub. No. 51, ASA, CSSA, SSSA, Madison, Wisconsin, pp. 193–218.

Rasmussen, P.E., H.P. Collins, and R.W. Smiley. 1989. Long-term management effects on soil productivity and crop yield in semi-arid regions of eastern Oregon. *Oregon Agr. Exp. Sta. Bull. 675.* 57 pp.

Rehm, G. 1992. Management of phosphate fertilizers in conservation tillage production systems. The midwest. *In* F.J. Sikora (Ed.). *Future Directions for Agricultural Phosphorus Research.* National Fert. and Environ. Res. Cent., TVA, Muscle Shoals, Alabama, 8 pp.

Reichle, D.E. 1977. The role of soil invertebrates in nutrient cycling. *In* U. Lohn and T. Persson (Eds.). *Soil Organisms as Components of Ecosystems.* Proc. 6th Intl. Coll. Soil Zool., Ecol. Bull. (Stockholm) 25:145–156.

Reicosky, D.C. 1983. Soil management for efficient water use. Soil-profile modifications effects on plant growth and yield in the southeastern United States. *In* H.M. Taylor, W.R. Jordan, and T.R. Sinclair (Eds.). *Limitations to Efficient Water Use in Crop Production.* Am. Soc. Agron., Madison, Wisconsin, pp. 471–477.

Romheld, V., and H. Marschner. 1983. Mechanism of iron uptake by peanut plants. I. Fe^3 reduction, chelate splitting, and release of phemolics. *Plant Physiol.* 71:949–954.

Russell, E.W. 1973. *Soil Conditions and Plant Growth.* Longman Press, London, England.

Russelle, M.P., and W.L. Hargrove. 1989. Cropping systems. Ecology and management. *In* R.F. Follett (Ed.). *Nitrogen Management and Ground Water Protection. Developments in Agricultural and Managed-Forest Ecology, 21*, Elsevier Press, Amsterdam, The Netherlands, pp. 277–317.

Saffigna, P.G., and D.R. Keeney, 1977. Nitrogen and chloride uptake by irrigated Russet Burbank potatoes. *Agron. J.* 69:258–264.

Salih, H.M., A.I. Yahya, A.M. Abdul-Raheui, and B.H. Munam. 1989. Availability of phosphorus in a calcareous soil treated with rock phosphate or superphosphate as affected by phosphate-dissolving fungi. *Plant Soil* 120: 181–185.

Sanchez, P.A., W. Conto, and S.W. Buol. 1982. The fertility capability soil classification system. Interpretation, applicability and modification. *Geoderma* 27:283–309.

Satchell, J.E. 1967. Lumbricidae. *In* A. Burges and F. Raw (Eds.). *Soil Biology.* Academic Press, London, England, pp. 259–322.

Saveson, I.L., Z.F. Lund, and L.W. Sloane. 1961. Deep tillage investigations of compacted soil in the cotton area of Louisiana. *USDA-ARS Pub. No. 41-19*, U.S. Govt. Print. Off., Washington, D.C.

Schenk, M.K., and S.A. Barber. 1979a. Phosphate uptake by corn as affected by soil characteristics and root morphology. *Soil Sci. Soc. Am. J.* 43:880–883.

Schenk, M.K., and S.A. Barber. 1979b. Root characteristics of corn genotypes as related to P uptake. *Agron. J.* 71:921–924.

Schepers, J.S. 1988. Role of cropping systems in environmental quality. Groundwater nitrogen. *In* W.L. Hargrove et al. (Eds.). *Cropping Strategies for Efficient Use of Water and Nitrogen.* Am. Soc. Agron. Publ. No. 51, Madison, Wisconsin, pp. 167–178.

Schepers, J.S., and F.E. Below. 1988. Influence of corn hybrids on nitrogen uptake and utilization efficiency. *Proc. 42nd Annual Corn Sorghum Industry Res. Conf.* Chicago, Illinois, pp. 172–186.

Schepers, J.S., and R.H. Fox. 1989. Estimation of N budgets for crops. *In* R.F. Follett (Ed.). *Nitrogen Management and Ground Water Protection. Development in Agricultural and Managed-Forest Ecology, 21.* Elsevier Press, Amsterdam, The Netherlands, pp. 221–246.

Schepers, J.S., and R. Saint-Fort. 1988. Comparison of potentially mineralized nitrogen using electroultrafiltration and four other procedures. *Proc. 3rd Intl. Electroultrafiltration Symposium.* Mannheim, Germany, Vol. 2:441–450.

Scheu, S. 1987. Microbial activity and nutrient dynamics in earthworm casts (Lumbricidae). *Biol. Fert. Soils* 5:230–234.

Schumacher, T.E., and A.J.M. Smucker. 1984. Effect of localized amoxia on *Phaseolus vulgaris* L. root growth. *J. Exp. Bot.* 35:1039–1047.

Schuman, G.E., and L.F. Elliott. 1978. Cropping an abandoned feedlot to prevent deep percolation of nitrate-nitrogen. *Soil Sci.* 126:237–243.

Scott, R.R. 1977. *Plant Root Systems. Their Function and Interaction with the Soil.* Book 6, McGraw-Hill, London, England. 298 pp.

Scotter, D.R., and P. Kanchanasut. 1981. Anion movement in a soil under pasture. *Aust. J. Soil Res.* 19:299–307.

Seligman, N.G., and H. Van Keulen. 1981. PAPRAN: A simulation model of annual pasture production limited by rainfall and nitrogen. *In* M.J. Frissel and J.A. van Veen (Eds.). *Simulation of Nitrogen Behaviour of Soil-Plant Systems.* Centre for Agric. Pub. Documentation, Wageningen, The Netherlands, pp. 192–221.

Sharpley, A.N. 1985. Phosphorus cycling in unfertilized and fertilized agricultural soils. *Soil Sci. Soc. Am. J.* 49:905–911.

Sharpley, A.N., and S.J. Smith. 1989. Mineralization and leaching of phosphorus from soil incubated with surface-applied and incorporated crop residues. *J. Environ. Qual.* 18:101–105.

Sharpley, A.N., and J.K. Syers. 1977. Seasonal variation in casting activity and in the amounts and release to solution of phosphorus forms in earthworm casts. *Soil Biol. Biochem.* 9:227–231.

Sharpley, A.N., J.K. Syers, and J.A. Springett. 1970. Effect of surface-casting earthworms on the transport of phosphorus and nitrogen in surface runoff from pasture. *Soil Biol. Biochem.* 11:459–462.

Sharpley, A.N., C.A. Jones, C. Gray, and C.V. Cole. 1984. A simplified soil and plant phosphorus model. II. Prediction of labile, organic, and sorbed phosphorus. *Soil Sci. Soc. Am. J.* 48:805–809.

Shipley, P.R., J.J. Meisinger, and A.M. Decker. 1992. Conserving residual corn fertilizer nitrogen with winter cover crops. (*Agron. J.*, in press).

Silberbush, M., and S.A. Barber. 1983. Prediction of phosphorus and potassium uptake by soybean with a mechanistic mathematical model. *Soil Sci. Soc. Am. J.* 47:262–265.

Singh, B., and G.J. Sekhon. 1977. Some measures of reducing leaching loss of nitrates beyond potential rooting zones. III. Proper crop rotations. *Plant Soil* 47:585–591.

Singh, S.S. 1984. Increase in neutral salt extractable cation exchange capacity of some acid soils as affected by $CaSO_4$ applications. *Can. J. Soil Sci.* 64:153–161.

Smika, D.E., and P.H. Grabouski. 1976. Anhydrous ammonia applications during fallow for winter wheat production. *Agron. J.* 68:919–922.

Smith, K.A. 1981. A model of denitrification in aggregated soils. *In* M.J. Frissel and J.A. van Veen (Eds.). *Simulation of Nitrogen Behavior of Soil-Plant Systems.* Centre for Agric. Pub. and Documentation, Wageningen, The Netherlands, pp. 259–266.

Smith, S.E., B.J. St. John, F.A. Smith, and D.J.D. Nicholas. 1985. Activity of glutamine synthetase and glutamate dehydrogenase in *Trifolium subterraneum* L. and *Allium cepa* L. Effect of mycorrhizal infection and phosphate nutrition. *New Phytol.* 99:211–217.

Smith, S.J., and A.N. Sharpley. 1990. Soil nitrogen mineralization in the presence of surface and incorporated crop residues. *Agron. J.* 82:112–116.

Smith, S.J., L.B. Young, and G.E. Miller. 1977. Evaluation of soil nitrogen mineralization protentials under modified field conditions. *Soil Sci. Soc. Am. J.* 41:74–76.

Smyth, T.J., and P.A. Sanchez. 1982. Phosphate rock and superphosphate combinations for soybeans in a Cerrado oxisol. *Agron. J.* 74:730–735.

Soil Survey Staff. 1975. Soil taxonomy. A basic system of soil classification for making and interpreting soil survey. *USDA-SCS Handbook No. 436*. U.S. Govt. Print. Off., Washington, D.C. 754 pp.

Sojka, R.E., D.L. Karlen, and E.J. Sadler. 1988. Planting geometrics and the efficient use of water and nitrogen. *In* W.C. Hargrove (Ed.). *Cropping Systems for Efficient Use of Water and Nitrogen*. Am. Soc. Agron. Spec. Pub. No. 51, ASA, CSSA, SSSA, Madison, Wisconsin, pp. 43–68.

Soper, R.J., and Y.P. Kalra. 1969. Effect of mode of application and source of fertilizer on phosphorus utilization by buckwheat, rape, oats, and flax. *Can. J. Soil. Sci.* 49:319–326.

Springett, J.A. 1983. Effect of five species of earthworms on some soil properties. *J. Appl. Ecol.* 20:865–872.

Springett, J.A., and J.K. Syers. 1979. The effect of earthworm casts on ryegrass seedlings. *In* T.K. Crosby and R.P. Pottinger (Eds.). *Proc. 2nd Australasian Conf. Grassl. Invert. Ecol.* Government Printer, Wellington, New Zealand, pp. 44–47.

Srivastava, O.P., and B.C. Sethi. 1981. Contribution of farm yard manure on the buildup of available zinc in an Aridisol. *Commun. Soil Sci. Plant Anal.* 12:355–361.

Stanford, G., and S.J. Smith. 1972. Nitrogen mineralization potentials of soils. *Soil Sci. Soc. Am. Proc.* 36:465–472.

Stecker, J.A., D.H. Sander, F.N. Anderson, and G.A. Peterson. 1988. Phosphorus fertilizer placement and tillage in a wheat-fallow cropping sequence. *Soil Sci. Soc. Am. J.* 52:1063–1068.

Stewart, J.W.B., and A.N. Sharpley. 1987. Controls on dynamics of soil and fertilizer phosphorus and sulfur. *In* R.F. Follett, J.W.B. Stewart, and C.V. Cole (Eds.). *Soil Fertility and Organic Matter as Critical Components of Production Systems*. Soil Sci. Soc. Am. Spec. Pub. 19, Madison, Wisconsin, pp. 101–121.

Stockdill, S.M.J. 1982. Effects of introduced earthworms on the productivity of New Zealand pastures. *Pedobiologia* 24:29–35.

Stockdill, S.M.J., and G.G. Cossins. 1966. The role of earthworms in pasture production and moisture conservation. *Proc. New Zeal. Grassland Assoc.* 28:168–183.

Stolzy, L.H., and K.P. Barley. 1968. Mechanical resistance encountered by roots entering compact soils. *Soil Sci.* 105:297–301.

Stolzy, L.H., J. Letey, T.E. Szuszkiewicz, and O.R. Lunt. 1961. Root growth and diffusion rates as functions of oxygen concentration. *Soil Sci. Soc. Am. Proc.* 25:463–467.

Stone, J.A. 1985. Poorly drained conditions and root development of eight indeterminate soybean cultivars. *Agron. J.* 77:751–789.

Strong, W.M., and R.J. Soper. 1974. Utilization of pelleted phosphorus by flax, wheat, rape, and buckwheat from a calcareous soil. *Agron. J.* 65: 18–21.

Sugi, Y., and M. Tanaka. 1978. Number and biomass of earthworm populations. *In* T. Kira, Y. Ono, and T. Hosokawa (Eds.). *Biological Production in a Warm-Temperature Evergreen Oak Forest of Japan*. J.I.B.P. Synthesis Vol. 18, University of Tokyo Press, Tokyo, pp. 171–178.

Sumner, M.E., H. Shahandah, J. Bouton, and J. Hammel. 1986. Amelioration of an acid soil profile through deep liming and surface application of gypsum. *Soil Sci. Soc. Am. J.* 50:1254–1258.

Syers, J.K., A.N. Sharpley, and D.R. Keeney. 1979. Cycling of nitrogen by surface-casting earthworms in a pasture ecosystem. *Soil Biol. Biochem.* 11: 181–185.

Tabatabai, M.A. 1984. Importance of sulphur in crop production. *Biogeochem.* 1:45–62.

Tabatabai, M.A. (Ed.) 1986. *Sulfur in Agriculture*. Agronomy 27, Am. Soc. Agron., Madison, Wisconsin.

Tanaka, D.L., and J.K. Aase. 1989. Influence of topsoil removal and fertilizer application on spring wheat yields. *Soil Sci. Soc. Am. J.* 53:228–232.

Taylor, H.M. 1983. Managing root systems for efficient water use. An overview. *In* H.M. Taylor, W.R. Jordan, and T.R. Sinclair (Eds.). *Limitations of Efficient Water Use in Crop Productivity*. Am. Soc. Agron., Madison, Wisconsin, pp. 87–113.

Taylor, H.M., and L.F. Ratliff. 1969. Root growth pressures of cotton, peas and peanuts. *Agron. J.* 61:398–402.

Thompson, L.F. 1990. A new rock phosphate. Is there a place for it in our agriculture? p. 8. *Am. Soc. Agron. Abstract*. Southern Section Meeting, Little Rock, Arkansas, February, 1990.

Tinker, P.B. 1978. Effects of vesicular-arbuscular mycorrhizae on plant growth. *Physiol. Veg.* 16:743–751.

Tisdall, J.M. 1978. Ecology of earthworms in irrigated watersheds. *In* W.W. Emerson, R.D. Bond, and A.R. Deater (Eds.). *Modification of Soil Structure*. John Wiley and Sons, Chichester, England, pp. 297–303.

Tiwari, K.N. 1979. Efficiency of Mussoorie rock phosphate, pyrites, and their mixtures in some soils of Uttar Pradesh. *Indian Soil Sci. Soc. Bull.* 12:519–526.

Tiwari, S.C., B.K. Tiwari, and R.R. Mischra. 1989. Microbial populations, enzyme activities, and nitrogen-phosphorus-potassium enrichment in earthworm casts and in surrounding soil of a pineapple plantation. *Biol. Fer. Soils* 8:178–182.

Torres, B.C., and F.T. Bingham. 1973. Salt tolerance of Mexican wheat. I. Effect of NO_3 and NaCl on mineral nutrition, growth, and grain production of four wheats. *Soil Sci. Soc. Am. Proc.* 37:711–715.

Treeby, M.T., and R.F.M. van Steveninck. 1988. Effects of salinity and phosphate on ion distribution in lupin leaflets. *Physiol. Plant* 73:317–322.

Treeby, M.T., H. Marschner, and V. Romheld. 1989. Mobilization of iron and other micronutrient cations from a calcareous soil by plant-borne, microbial, and synthetic metal chelators. *Plant Soil* 114:217–226.

Triplett, G.B. Jr., and D.M. Van Doren. 1969. Nitrogen phosphorus and potassium fertilization of non-tilled maize. *Agron. J.* 61:637–639.

Tucker, B.B., and L.W. Murdock. 1984. Nitrogen use in the southcentral states. *In* Hauck, R.D. (Ed.). *Nitrogen in Crop Production*. Am. Soc. Agron., Madison, Wisconsin, pp. 735–744.

Ulmer, S.E. 1979. *Aluminum Toxicity and Root DNA Synthesis in Wheat*. Ph.D Dissertation, Iowa State University, Ames.

Unger, P.W., H.V. Eck, and J.T. Musick. 1981. Alleviating plant water stress. *In* G.F. Arkin and H.M. Taylor (Eds.). *Modifying the Root Environment to*

Reduce Crop Stress. Am. Soc. Agric. Eng. Monogr. No. 4, St. Joseph, Michigan, pp. 61–96.

Van Doren, D.M., and J.L. Haynes. 1961. Subsoil tillage of fragipan soils of Ohio. *Ohio Agric. Exp. Stn. Res. Civ. 95*, Columbus. 4 pp.

Van Schilfgaarde, J. 1974. Drainage for Agriculture. Am. Soc. Agron. Monogr. No. 17, Madison, Wisconsin. 700 pp.

Veihmeyer, F.J., and A.H. Hendrickson. 1948. Soil density and root penetration. *Soil Sci.* 65:487–493.

Wagenet, R.J., R.R. Rodriguez, W.F. Campbell, and D.L. Turner. 1983. Fertilizer and salty water effects on *Phaselous*. *Agron. J.* 75:161–166.

Wagger, M.G., and D.B. Mengel. 1988. The role of nonleguminous cover crops in the efficient use of water and nitrogen. *In* W.L. Hargrove (Ed.). *Cropping Strategies for Efficient Use of Water and Nitrogen*. Am. Soc. Agron. Spec. Pub. No. 51, ASA, CSSA, SSSA, Madison, Wisconsin, pp. 115–127.

Walker, D.R., and G. Doornenbal. 1972. Soil sulfate. II. As an index of the sulfate available to legumes. *Can. J. Soil Sci.* 52:261–266.

Watson, D.H., and J.R. Ellis. 1990. Indigenous VA mycorrhizal fungi of an agronomic soil in eastern Nebraska. *In 8th NACOM Abstracts on Innovation and Hierarchial Integration*, University of Wyoming, Cheyenne, pp. 304.

Weaver, J.E. 1926. *Root Development of Field Crops*. McGraw-Hill Book Co., New York.

Welch, L.F., D.L. Mulvaney, L.V. Boone, G.E. Mikibben, and J.W. Pendleton. 1966. Relative efficiency of broadcast versus banded phosphorus for corn. *Agron. J.* 58:283–287.

White, R., M.J. Hedley, N.S. Bolan, and P.E.H. Gregg. 1989. Recent developments in the use of phosphate fertilizers on New Zealand pastures. *J. Aust. Inst. Agric. Sci.* 2:26–32.

Wilhelm, W.W., L.N. Mielke, and C.R. Fenster. 1982. Root development of winter wheat as related to tillage in western Nebraska. *Agron. J.* 74:85–88.

Williams, J.R. 1983. The physical components of the EPIC model. *In* S.A. El-Swaify, W.C. Moldenhauer, and A. Lo (Eds.). *Soil Erosion and Conservation*. Soil Conserv. Soc. Am., Ankeny, Iowa, pp. 272–284.

Williams, J.R., C.A. Jones, and P.T. Dyke. 1984. A modeling approach to determining the relationship between erosion and soil productivity. *Trans. Am. Soc. Agric. Eng.* 27:129–144.

Williamson, R.E. 1964. The effect of root aeration on plant growth. *Soil Sci. Soc. Am. Proc.* 28:86–90.

Yang, J.E., E.O. Skogley, S.J. Georgitis, B.E. Schaff, and A.H. Ferguson. 1991. Phytoavailability soil test: Development and verification of theory. *Soil Sci. Soc. Am. J.* 55:1358–1365.

Yao, J., and S.A. Barber. 1986. Effect of one phosphorus rate placed in different soil volumes on P uptake and growth of wheat. *Commun. Soil Sci. Plant Anal.* 17:819–827.

Yost, R.S., and R.L. Fox. 1979. Contribution of mycorrhizae to P nutrition of crops growing on an oxisol. *Agron. J.* 71:903–908.

Zuzel, J.F., J.L. Pikul, Jr., and P.E. Rasmussen. 1990. Tillage and fertilizer effects on water infiltration. *Soil Sci. Soc. Am. J.* 54:205–208.

Index

A

A. longa, 190–191
Abscisic acid, 17
Actinomycetes, 173
Aeration, 1
Aerobic conditions, 35
Aeroponics, 35
Africa, 115
Aggregates, 75–77
  size, 30–31
Aggregation, 155, 183
Alabama, 189
Alfalfa, 9, 75, 99–100, 112–113, 118, 121–122, 155, 185, 188
Allium cepa, 4, 170
Allolobophlora caliginosa, 177–178, 190–191
Amaranthus, 125
Amaranthus sp., 125
Amino acids, 15, 98, 116
Ammonia, volatilization, *see also* nitrogen, 192
Anacardium occidentale L., 112
Anaerobic conditions, 34
Apoplasmic flow, 54, 55
Appling soil, 121
Arachis hypogaea L., 109, 155
Artemisia tridentata, 61
Australia, 110, 112, 120, 195
Available water, 154
Avena sativa, 90, 155

B

Bacteria, 15, 128, 173
  nitrogen fixing, 161
Bahiagrass, 189
Bamboo, 114
Bambusa multiplex, 114
Barley, 4–8, 36, 55, 75, 99, 104, 110, 114, 119, 170, 177
Barnyard grass, 114
Bean, 6, 9, 12, 99, 114, 153, 161, 170
Beets, 169
Bermuda grass, 61, 112, 157
Beta vulgaris L., 11, 166, 169
Betula papyrifera, 128
Biological activity, 15
Birdfoot trefoil, 158
Bladen soil, 157
Bothriochloa intermedia, 125
Brassica napus L., 119, 188
Brassica oleraceae L., 58, 112, 169
Brazil, 104, 110, 113, 117, 122
Broccoli, 58, 112, 169
Buffer capacity, 196
Bulk density, 15, 75–76, 79–80, 91

C

C:N ratio, 17, 159, 175
Cabbage, 169–170
Cahaba loamy sand, 189

Calcium, deficiency, 100, 105, 108–111, 113, 122, 129, 156
Canada, 33
Cancer, esophageal, 115
Carbon dioxide, 34, 47
  levels, 35, 37
  rhizosphere, 36
Carrot, 9, 14, 170
Carthamus tinctorius L., 119, 186
Cashew, 112
Casparian Strip, 55
Cation exchange capacity, 101–102, 106, 111, 116
Cattle dung, 174
Cauliflower, 112
Cecil soil, 118
Centrosema pubescens, 105
Chelates, 16
Chickpea, 13, 119
Chlorine, 169
Cicer arietinum L., 13, 119
Citrus, 112
Citrus aurantium L., 128
Citrus sinensis, 128
Clover, 9, 112, 164
Cocksfoot, 112
Cohesion theory, 53
Cold tolerance, 34
Comly silt loam, 40
Compaction, 3, 9, 30–31, 74, 82–83, 86, 91–92, 154–155
  subsoil, 88–89, 91
Compost, animal *see also* Manure, 151
Corn *see also* Maize, 1, 32–34, 37, 39–40, 153–155, 157–158, 164, 169–170, 184–185, 187–189
Cotton, 1, 9–10, 14, 61, 74, 98, 101, 104, 115, 118, 121, 125, 155–157, 169, 189
Cover crops, 186–187, 199
Cowpea, 13, 100, 127
Crimson clover, 188
Crop residue, 154, 158–159, 166–167, 174, 183, 198
  decay, 17
Crop rotation, 162, 199
Cropping practices, 74
Cropping systems, 152, 185
Cucumber, 114
Cucumus sativus L., 114
Cyamopsis tetragonoloba (L.) Taub, 13
Cynodon dactylon L., 61, 112, 157

D

Dactylis glomerata L., 112
Darcian flow, 61
Daucus carota, 170
Davidson clay loam, 121
Denitrification, 192
Dickson soil, 157
Diffusion, 36–37, 56, 179, 196
  rates, 35
Diffusive resistance, 56
Disease
  incidence, 28
  resistance, 28
Douglas fir, 105
Drainage, 62, 153–154, 185, 187
Drought, 98, 103, 119–120
  tolerance, 129, 188

E

Earthworms, 152, 173–178, 190–191
England, 174
Environmental quality, 152
Environments
  arid, 158
  semi-arid, 158
Equisetum arvense L., 114
Equisetum havemale L., 114
Erosion, 182
  losses, 183
  wind, 183
Euphorbia pulcherrina, 125
European beech, 105
Evaporation, losses, 159
Evapotranspiration, 152, 186

F

Fagus sylvatica L., 105
Fallow, 185
Feedlots, abandoned, 185
Fernbrake, 114

Fertilizer, 129, 151–152, 168
  application, 74
  availability index, 195
  leaching, 165
  micronutrients, 158, 172
  nitrogen, 5–6, 99, 117, 130, 154, 160, 165, 169–170, 186–187
  phosphorus, 117, 120, 123, 161, 163–165, 169, 171, 181–182
  placement, 163, 193
    deep placement, 117, 120
  potassium, 117
  residual, 166, 183, 188
  starter, 164
  sulfur, 165
  timing, 164
  use-efficiency, 165, 167, 172, 178
Fescue, 157
Festuca elatior, 157
Flax, 107
Florida, 114, 162
Fragipan, 155
Fruit crops, 166
Fungi, 15, 63, 125, 128, 161

G

Georgia, 108, 121
Germination, 188
  seed, 32–33, 116, 124
Global climate change, 36, 66
Glomus fasciculatus, 128
Glycine max L., 9–10, 33, 58–59, 78, 99, 153
Gossypium hirsutum L., 1, 9–10, 61, 74, 98, 155
Grass barriers, 183
Grasslands, 17
Green Revolution, 28
Growing degree days, 8
Guar, 13
Gypsum *see also* Phosphogypsum, 116–118, 121–124, 157

H

Hapludult, 121
Hawaii, 108
Helianthus annuus L., 14, 58, 104
Hemarthia altissima, 125
Hibiscus cannabinus L., 186
Histosols, 114
Hordeum distichum L., 8
Hordeum vulgare L., 4, 36, 55, 75, 99, 170, 177
Hormones, 17, 154, 174
Horsetail, 114
Humic material, 114
Hydraulic conductivity, 15, 60, 90, 104
Hydroponics, 35, 74

I

Indolaecetic acid, 18
Infiltration, 154, 168, 183
Insect resistance, 28
Iowa, 33
Iran, 115
Iron, deficiency, 100, 107, 158
Irrigation, 3, 64, 129, 152, 168, 189
  water, 114
  alternate flow, 61
  frequency, 168
  surge, 168
  systems, 168

J

Jack pine, 128
Japan, 174

K

Kansas, 163
Kaolinite, 121
Kenaf, 186

L

L. corniculatus, 158
L. rubellus, 178, 190–191
Lactuca sativa L., 10, 108
Latasols, 114, 122
Leaching, 177, 198
  fraction, 154
Lespedeza, 100
Lespedeza stipulacea, 100

Lettuce, 10, 108
Leucaena, 109
Leucaena leucocephala L., 109
Lime, 108–109, 111–112, 117, 129–130, 151, 157, 190
  deep placement, 119
Limpograss, 125
Lintonia soil, 118
Linum Usitaissimum L., 107
Lolium multiflorum L., 124
Lolium perenne L., 5, 7, 161, 177
Lotus peduncalatus, 127
Lumbricus rubellus, 177
Lumbricus terrestris, 174
Lupin, 64, 119
Lupinus sp., 64, 119
Lycopersicon lycopersicum, 12, 58, 170
Lycopersicon esculentum L., 10, 114
Lysimeter, 66

M

Macroorganisms, 37
Macropores, 77
Magnesium, deficiency, 100, 110–111, 113, 156
Maize *see also* Corn, 2–3, 6–7, 55, 64, 74, 78–83, 85–88, 90–91, 105, 114, 125, 159
Manganese, deficiency, 112
Manure, 152
  animal, 123
  green, 123, 130
  management, 160
Marsh trefoil, 127
Mechanical resistance, 65, 90
Medicago sativa L., 75, 99, 155
Mediterranean grasses, 61
Melilotus sp., 100
Mesquite, 61
Microorganisms, 35, 37, 98, 129, 131, 159
Micropores, 77
Millet, 170
Minespoils, 163
Minnesota, 80, 87, 89–90
Mitscherlich curve, 195
Mixed cropping, 188
Models
  Century, 194, 196
  Classen-Barber, 196
  crop, 198
  Decide, 195–196
  development, 199
  EPIC, 196–197
  mathematical, 192
  simulation, 192
  WHTROOT, 8
Molybdenum, deficiency, 108, 113, 156
Montmorillonite, 39
Mulch, straw, 33
Multiple cropping, 186
Mung bean, 127
Mycorrhizae, 16, 63, 128, 152, 161, 173, 178–180

N

^{15}N, 187
Nebraska, 164
New York, 32–33, 42
New Zealand, 162
Nicotiana plumbaginifolia, 125
Nicotiana tabacum L., 108
Nigeria, 110, 115
Nirogen, rates, 118
Nitrate, 31, 116, 175
  concentrations, 173
  leaching, 179
  losses, 188
  pollution, 186
  residual, 165
  under alfalfa, 186
Nitrification inhibitor, 167
Nitrogen *see also* Fertilizer, 14, 152, 189
  ammonium, 31, 116, 158, 180
  anhydrous ammonia, 160
  deficiency, 58
  denitrification, 160
  fertility, 7
  fixation, 113, 160
  immobilization, 159–160
  leaching, 167
  leaching losses, 160
  metabolism, 116
  mineralization, 165

nitrate, 31, 116
placement, 163
slow release fertilizers, 160
sources, 158–159
supply, 5
tissue, 169
urea, 160
volatilization, 160
Norfolk soil, 155, 157
North Carolina, 161–162
Norway spruce, 105, 128
Nut crops, 166
Nutrients
availability, 199
cycling, 167, 173, 176, 192–193
essential, 97
leaching, 175
losses, 154
mineralization, 194
transport, 53

O

Oat, 90, 114, 119, 155, 189
Octolasion cyanenum, 174
Ohio, 112, 155
Ohm's law, 57
Old World bluestems, 125
Onion, 4, 170
Organic matter, 17–18, 31, 35–36, 74, 108, 123, 152, 158, 163, 176–177, 180, 194, 198
cycling, 16, 199
decomposition, 152, 194
loss, 155
management, 158
sources, 160
waste, 160
Oryza sativa L., 2, 100
Osmotic potential, 60, 64–65
Oxisols, 110–111, 116–117
Oxygen deficiency, 58

P

^{32}P, 177
Paleudult, 155
Pan layers, 30–31
Panicum virgatum L., 102
Paper birch, 128
Paspalum notatum, 189
Pathogens, 154
Pea, 10, 119
Peanut, 109, 127, 155–157
Pearl millet, 3
Penetrometer resistance, 75, 80–83, 91, 156
Penicillium bilati, 181
Pennisetum typhoides, 3
pH, 102, 107, 109–110, 112–113, 122, 127, 156, 158, 161–162, 180
Phalaris, 7, 115
Phalaris aquatica L., 7
Phalaris tuberosa, 61
Phaseolus vulgaris L., 9, 13, 99, 105, 153, 161
Pheretima sp., 174
Phosphogypsum *see also* Gypsum, 116, 122
Phosphorus *see also* Fertilizer, 76, 152, 166, 170, 183
availability, 114
cycling, 177
deep placement, 120
deficiency, 7–8, 100, 113
efficiency, 115
fixation, 130
mineralization, 194
nutrition, 13
Olsen test, 166, 177
organic, 192
placement, 163
plant concentration, 170
retention test, 198
rock phosphate, 131, 161, 163, 191
Photosynthesis, 15
Physical
conditions, 92
properties, 83, 88
resistance, 73
restriction, 74
Picea abies L., 105
Picea rubens, 106
Pinus banksiana L., 128
Pinus radiata, 13
Pisum sativum L., 9, 11, 119, 126

Plant
  auxins, 178
  stature, 28
Plow layer, 33–34, 97
Plowpans, 120, 155, 189
Poinsettia, 125
Pore size, 37
Potassium, 121
  deficiency, 7
Potato, 13, 120
Precipitation, use-efficiency, 185
Prosopis tamarugo, 61
Pseudotsuga menziesii, 105
Pteridium aquilinum L., 114
Puerto Rico, 117

Q
Quercus, 64

R
Radish, 14
Radon gas, 122
Rain-fed conditions, 88
Rape, 188
Rapeseed, 119
Red clover, 99
Red spruce, 106
Rhizobia, 98
  soybean, 100
Rhizobium japonicum, 126
Rhizobium leguminosarum, 126
Rhizobium lupini, 125
Rhizobium meliloti, 127
Rhizobium phaseoli, 126–127
Rhizobium trifoli, 127
Rhizosphere, 15, 30–31, 35–36, 65, 196
Rhizotron, 10
Rice, 2, 100, 103, 110, 114–115, 169–170
Root cortex, 54
Root penetration, 157
Root zone, 62, 151, 165
  temperature, 32–33
Roots
  adventitious, 2, 34, 36
  coronal, 2
  crown, 2, 5
  nodal, 7, 34, 36
  seminal, 2, 3, 7
  temperature, 58
Rooting depth, 39, 89
Rye, 104, 187
Ryegrass, 5, 7, 115, 124, 161, 177–178, 191

S
Saccharum officinarium L., 114
Safflower, 119, 185–186
Sagebrush, 61
Salinity *see also* Soil, 153–154, 169–171
  genotype tolerance, 125
Scouring rush, 114
Secale cereale L., 104
Seedling vigor, 1
Sesame, 170
Sesamum indicum, 170
Sewage sludge, 108
Shelter belts, 183
Silicon, deficiency, 97, 114
Snapbean, 105
Soil
  acidic subsoil, 98, 111
  acidity, 15, 158
  aeration, 189
  amendments, 151, 160
  biota, 16
  bulk density, 155
  cracks, 15
  density, 30
  fauna, 173, 199
  fertility, 1, 6, 16, 28, 151, 156, 166, 178, 183–184, 195
  moisture, 3
  physical properties, 15
  productivity, 27, 182, 184
  properties, 2
  salinity, 58, 97, 116, 124, 152
  strength, 3, 15, 65
  structure, 15, 173
  subsoil acidity, 157
  sustainability, 27

temperature, 1, 3, 6, 15, 31–33, 47, 154, 174, 192, 195
tilth, 198
water
availability, 64
content, 15, 89, 151, 195
potential, 61
Solanum tuberosum L., 13, 120
Sorghum, 2, 5, 99–100, 104, 108, 157, 163, 170, 188
Sorghum bicolor L., 2, 99, 157
South Africa, 119, 122
South Carolina, 112
Soybean, 9–10, 12–14, 33, 36, 42, 58–59, 78–80, 83–85, 87–88, 91, 99, 105, 109, 112, 114, 153, 158–159
Soybeans, 9, 42
Spatial variability, 37, 40, 43, 45, 178
Stress
aluminum, 104
drought, 100, 104
heat-induced, 32
mineral, 124
physical, 34
salinity, 57
water, 9, 156, 169
Subclover, 102
Sugar beets, 166
Sugar cane, 74, 114
Sugarbeet, 11
Sulfur
deficiency, 166
mineralization, 166
organic, 166
Sunflower, 14, 58, 104
Sustainable agriculture, 130
Sweden, 188
Sweet clover, 100
Switchgrass, 102
Symplasmic flow, 54–55
Syria, 13

T
Tanzania, 112
Temperature *see also* Soil, 34
Termites, 173
Thailand, 110
Thallosphere, 30
Tillage, 33, 74, 86, 88, 151, 155, 159, 178, 189, 195, 199
alternative, 32
conservation, 91, 166, 183
conventional-till, 32
deep, 83
excess, 155
moldboard plowing, 80–81, 83–84
no-till, 32, 159, 183
primary, 80
reduced, 80, 82, 85, 157, 166, 181–182
rototilling, 83
secondary, 75
slit, 120
subsoiling, 83
techniques, 36
Tobacco, 108
Togo, 162
Tokomaru silt loam, 190
Tolerance, aluminum, 114
Tomato, 10, 12, 38, 58, 114, 169–170
Toxicity
aluminum, 97–98, 100, 102–103, 105, 109–113, 115, 117, 122, 127–131, 156
hydrogen ion, 98–99, 109
manganese, 97–98, 107–109, 112, 115, 129, 131, 156
Transpiration, 62
rate, 57
Trifolium incarnatum, 188
Trifolium pratense L., 99
Trifolium repens L., 125
Trifolium sp., 112
Trifolium subterraneum L., 102
Triticale, 119
Triticum aestivum L., 2, 65, 90, 98
Triticum vulgare, 28, 154
Tropics, 128
Turgor pressure, 1
Typic Chromoxererts, 39
Typic Fragidualf, 40
Typic Paleudult, 189
Typic Xerothents, 39

U
Udic Haplustoll, 80
Ultisols, 109, 115, 118
United States, 32–33, 74, 91, 104, 108, 120, 162, 186, 188, 197

V
Vermiculite, 10
Vicia faba L., 6, 11, 12
Vigna radiata, 127
Vigna unguicalata, 13, 100

W
Weed control, 75
Wheat, 2, 5, 28, 34, 45, 65, 90, 98, 100, 107, 110, 112–114, 119, 121, 124, 154, 161, 164, 169–170, 181, 184, 188
White clover, 115, 125
Wilting point, 59
Windbreaks, 183

X
X. tritosecale, 119

Y
Yellow leaf spot, 112
Yolo sandy loam, 39

Z
Zambia, 118
Zea mays L., 1–2, 33, 55, 64, 74, 105, 153
Zinc, deficiency, 158